Name_____

Laboratory Dos And Don'ts

Identify what is wrong in each laboratory activity.

1.

_____

_____

_____

4.

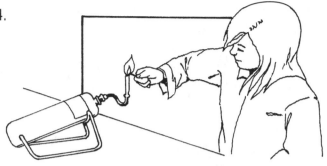

_____

_____

2.

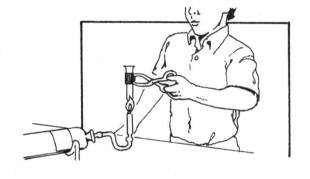

_____

_____

_____

5.

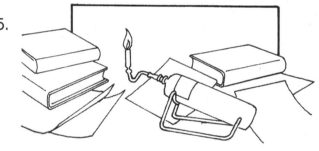

_____

_____

3.

_____

_____

_____

6.

_____

_____

Laboratory Equipment

Label the lab equipment.

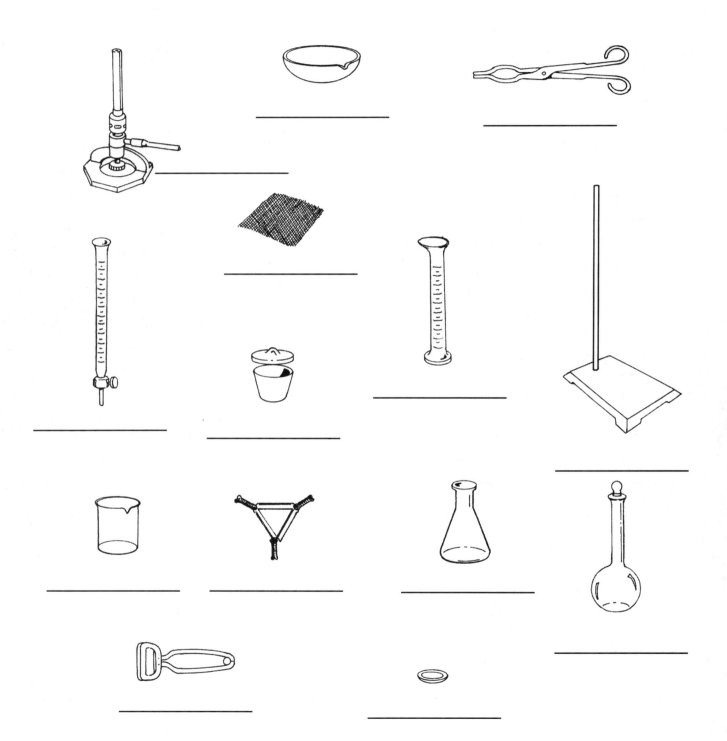

Triple And Four Beam Balances

Identify the mass on each balance.

Triple Beam Balance

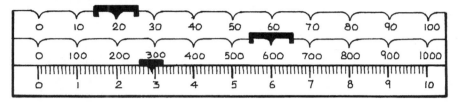

_____

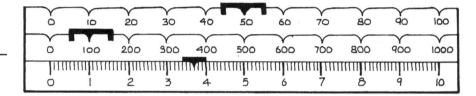

_____

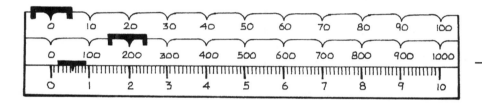

_____

Four Beam Balance

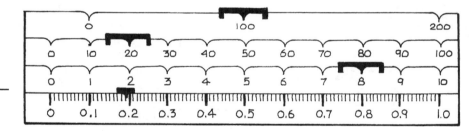

_____

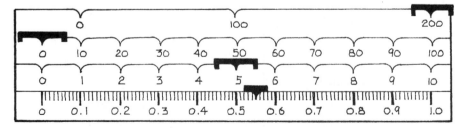

_____

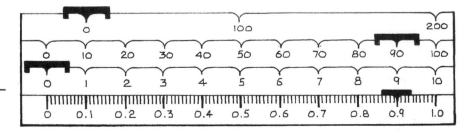

_____

Measuring Liquid Volume

Identify the volume indicated on each graduated cylinder. The unit of volume is mL.

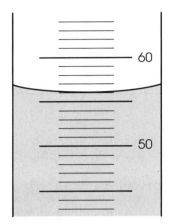

1. _____

2. _____

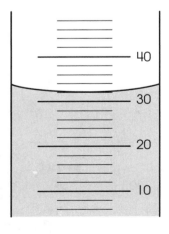

3. _____

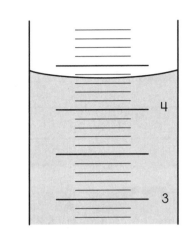

4. _____

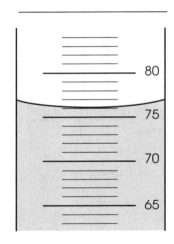

5. _____

6. _____

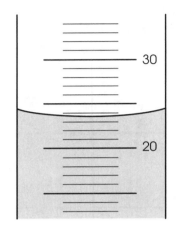

7. _____

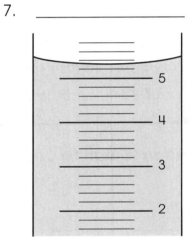

8. _____

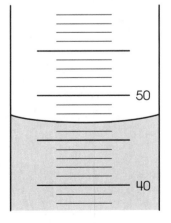

9. _____

Reading Thermometers

Identify the temperature indicated on each thermometer.

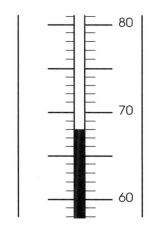

1. _____

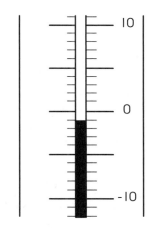

2. _____

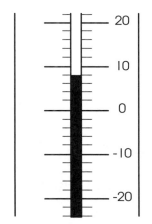

3. _____

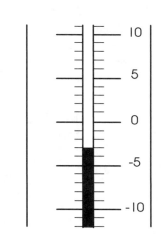

4. _____

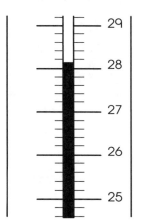

5. _____

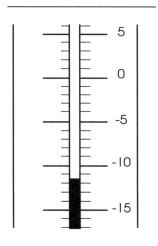

6. _____

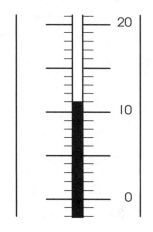

7. _____

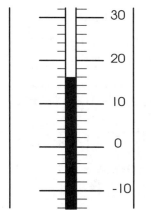

8. _____

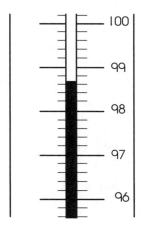

9. _____

Metrics And Measurements

In the chemistry classroom and lab, the metric system of measurement is used. It is important to be able to convert from one unit to another.

mega-	kilo-	hecto-	deca-	Basic Units	deci-	centi-	milli-	micro-
(M)	(k)	(h)	(da)	gram (g)	(d)	(c)	(m)	(μ)
1,000,000	1,000	100	10	liter (L)	0.1	0.01	0.001	0.000001
10^6	10^3	10^2	10^1	meter (m)	10^{-1}	10^{-2}	10^{-3}	10^{-6}

Unit Factor Method

1. Write the given number and unit.
2. Set up a conversion factor (fraction used to convert one unit to another).
    a. Place the given unit as the denominator of the conversion factor.
    b. Place the desired unit as the numerator.
    c. Place a one in front of the larger unit.
    d. Determine the number of smaller units needed to make one of the larger units.
3. Cancel the units. Solve the problem.

Example 1: 55 mm = _____ m

$$\frac{55 \text{ mm}}{} \left| \frac{1 \text{ m}}{1,000 \text{ mm}} \right. = 0.055 \text{ m}$$

Example 2: 88 km = _____ m

$$\frac{88 \text{ km}}{} \left| \frac{1,000 \text{ m}}{1 \text{ km}} \right. = 88,000 \text{ m}$$

Example 3: 7,000 cm = _____ hm

$$\frac{7,000 \text{ cm}}{} \left| \frac{1 \text{ m}}{100 \text{ cm}} \right| \frac{1 \text{ hm}}{10,000 \text{ cm}} = 0.7 \text{ hm}$$

Example 4: 8 daL = _____ dL

$$\frac{8 \text{ daL}}{} \left| \frac{10 \text{ L}}{1 \text{ daL}} \right| \frac{100 \text{ dL}}{1 \text{ daL}} = 800 \text{ dL}$$

The unit factor method can be used to solve virtually any problem involving changes in units. It is especially useful in making complex conversions dealing with concentrations and derived units.

Convert each measurement.

1. 35 mL = _____ dL

2. 275 mm = _____ cm

3. 1,000 mL = _____ L

4. 25 cm = _____ mm

5. 0.075 m = _____ cm

6. 950 g = _____ kg

7. 1,000 L = _____ kL

8. 4,500 mg = _____ g

9. 0.005 kg = _____ dag

10. 15 g = _____ mg

Dimensional Analysis (Unit Factor Method)

Using this method, it is possible to solve many problems by using the relationship of one unit to another. For example, 12 inches = one foot. Since these two numbers represent the same value, the fractions 12 in./1 ft. and 1 ft./12 in. are both equal to one. When you multiply another number by the number one, you do not change its value. However, you may change its unit.

Example 1: Convert 2 miles to inches.

$$2 \text{ miles} \times \frac{5,280 \text{ ft.}}{1 \text{ mile}} \times \frac{12 \text{ inches}}{1 \text{ ft.}} = 126,720 \text{ in.}$$

Example 2: How many seconds are in 4 days?

$$4 \text{ days} \times \frac{24 \text{ hrs.}}{1 \text{ day}} \times \frac{60 \text{ min.}}{1 \text{ hr.}} \times \frac{60 \text{ sec.}}{1 \text{ min.}} = 345,600 \text{ sec.}$$

Solve each problem. Round irrational numbers to the thousandths place.

1.   3 hr. = _____ sec.

2.   0.035 mg = _____ cg

3.   5.5 kg = _____ lb.

4.   2.5 yd. = _____ in.

5.   1.3 yr. = _____ hr.

6.   3 moles = _____ molecules (1 mole = 6.02×10^{23} molecules)

7. 2.5×10^{24} molecules = _____ moles

8.   5 moles = _____ liters (1 mole = 22.4 liters)

9.   100. liters = _____ moles

10.   50. liters = _____ molecules

11. 5.0×10^{24} molecules = _____ liters

12. 7.5×10^3 mL = _____ liters

Scientific Notation

Scientists very often deal with very small and very large numbers, which can lead to a lot of confusion when counting zeros. We can express these numbers as powers of 10.

Scientific notation takes the form of $M \times 10^n$ where $1 \le M < 10$ and n represents the number of decimal places to be moved. Positive n indicates the standard form is a large number. Negative n indicates a number between zero and one.

Example 1: Convert 1,500,000 to scientific notation.

Move the decimal point so that there is only one digit to its left, for a total of 6 places.

$$1,500,000 = 1.5 \times 10^6$$

Example 2: Convert 0.000025 to scientific notation.

For this, move the decimal point 5 places to the right.

$$0.000025 = 2.5 \times 10^{-5}$$

(Note that when a number starts out less than one, the exponent is always negative.)

Convert each number to scientific notation.

1. 0.005 = _____

2. 5,050 = _____

3. 0.0008 = _____

4. 1,000 = _____

5. 1,000,000 = _____

6. 0.25 = _____

7. 0.025 = _____

8. 0.0025 = _____

9. 500 = _____

10. 5,000 = _____

Convert each number to standard notation.

11. 1.5×10^3 = _____

12. 1.5×10^{-3} = _____

13. 3.75×10^{-2} = _____

14. 3.75×10^2 = _____

15. 2.2×10^5 = _____

16. 3.35×10^{-1} = _____

17. 1.2×10^{-4} = _____

18. 1×10^4 = _____

19. 1×10^{-1} = _____

20. 4×10^0 = _____

Name_____

Significant Figures

A measurement can only be as accurate and precise as the instrument that produced it. A scientist must be able to express the accuracy of a number, not just its numerical value. We can determine the accuracy of a number by the number of significant figures it contains.

1. All digits 1–9 inclusive are significant.

    Example: 129 has 3 significant figures.

2. Zeros between significant digits are always significant.

    Example: 5,007 has 4 significant figures.

3. Trailing zeros in a number are significant only if the number contains a decimal point. Sometimes, a decimal may be added without any number in the tenths place.

    Example: 100.0 has 4 significant figures.

    100. has 3 significant figures.

    100 has 1 significant figure.

4. Zeros in the beginning of a number whose only function is to place the decimal point are not significant.

    Example: 0.0025 has 2 significant figures.

5. Zeros following a decimal significant figure are significant.

    Example: 0.000470 has 3 significant figures.

    0.47000 has 5 significant figures.

Determine the number of significant figures in each number.

1. 0.02 _____
2. 0.020 _____
3. 501 _____
4. 501.0 _____
5. 5,000 _____

6. 5,000. _____
7. 6,051.00 _____
8. 0.0005 _____
9. 0.1020 _____
10. 10,001 _____

Determine the location of the last significant place value by placing a bar over the digit. (Example: 1.70$\bar{0}$)

11. 8,040
12. 0.90100
13. 3.01×10^{21}

14. 0.0300
15. 90,100
16. 0.000410

17. 699.5
18. 4.7×10^{-8}

19. 2.000×10^2
20. 10,800,000.0

Calculations Using Significant Figures

When multiplying and dividing, limit and round to the least number of significant figures in any of the factors.

> **Example 1:** 23.0 cm × 432 cm × 19 cm = 188,784 cm^3
>
> The answer is expressed as 190,000 cm^3 since 19 cm has only two significant figures.

When adding and subtracting, limit and round your answer to the least number of decimal places in any of the numbers that make up your answer.

> **Example 2:** 123.25 mL + 46.0 mL + 86.257 mL = 255.507 mL
>
> The answer is expressed as 255.5 mL since 46.0 mL has only one decimal place.

Perform each operation, expressing the answer in the correct number of significant figures.

1. 1.35 m × 2.467 m = _____

2. 1,035 m^2 ÷ 42 m = _____

3. 12.01 mL + 35.2 mL + 6 mL = _____

4. 55.46 g – 28.9 g = _____

5. 0.021 cm × 3.2 cm × 100.1 cm = _____

6. 0.15 cm + 1.15 cm + 2.051 cm = _____

7. 150 L^3 ÷ 4 L = _____

8. 505 kg – 450.25 kg = _____

9. 1.252 mm × 0.115 mm × 0.012 mm = _____

10. 1.278 × 10^3 m^2 ÷ 1.4267 × 10^2 m = _____

Percentage Error

Percentage error is a way for scientists to express how far off a laboratory value is from the commonly accepted value.

The formula is:

% error = absolute value → $\left| \dfrac{\text{Accepted Value} - \text{Experimental Value}}{\text{Accepted Value}} \right| \times 100$

Determine the percentage error in each problem.

1. Experimental value = 1.24 g
   Accepted value = 1.30 g

2. Experimental value = 1.24 x 10⁻² g
   Accepted value = 9.98 x 10⁻³ g

3. Experimental value = 252 mL
   Accepted value = 225 mL

4. Experimental value = 22.2 L
   Accepted value = 22.4 L

5. Experimental value = 125.2 mg
   Accepted value = 124.8 mg

Temperature and Its Measurement

Temperature (which measures average kinetic energy of the molecules) can be measured using three common scales: **Celsius**, **Kelvin**, and **Fahrenheit**. Use the following formulas to convert from one scale to another. Celsius is the scale most desirable for laboratory work. Kelvin represents the absolute scale. Fahrenheit is the old English scale, which is rarely used in laboratories.

$$°C = K - 273 \qquad\qquad K = °C + 273$$

$$°F = \frac{9}{5}\,°C + 32 \qquad\qquad °C = \frac{5}{9}\,(°F - 32)$$

Complete the chart. All measurements are good to 1°C or better.

	°C	K	°F
1.	0°C		
2.			212°F
3.		450 K	
4.			98.6°F
5.	-273°C		
6.		294 K	
7.			77°F
8.		225 K	
9.	-40°C		

Freezing and Boiling Point Graph

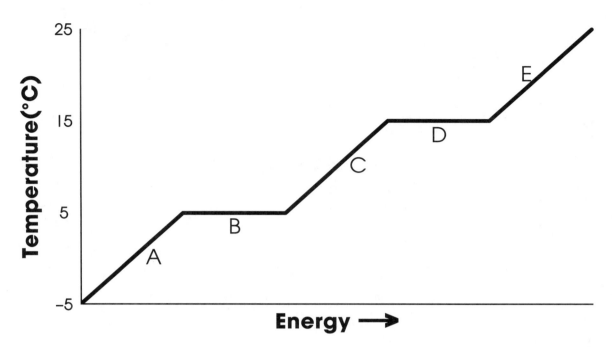

Use the graph to answer each question.

1. Which is the freezing point of the substance? _____

2. Which is the boiling point of the substance? _____

3. Which is the melting point of the substance? _____

4. Which letter represents the range where the solid is being warmed? _____

5. Which letter represents the range where the liquid is being warmed? _____

6. Which letter represents the range where the vapor is being warmed? _____

7. Which letter represents the melting of the solid? _____

8. Which letter represents the vaporization of the liquid? _____

9. Which letter(s) shows a change in potential energy? _____

10. Which letter(s) shows a change in kinetic energy? _____

11. Which letter represents condensation? _____

12. Which letter represents crystallization? _____

Phase Diagram

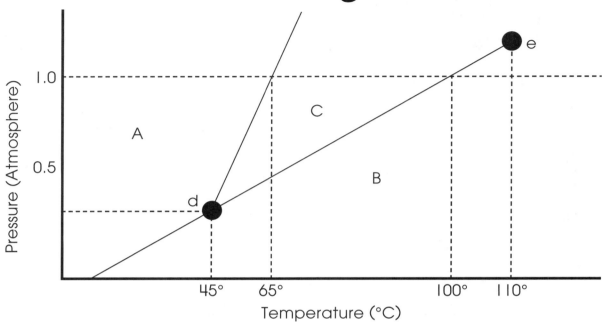

Use the diagram to answer each question.

1. Which **section** represents the solid phase? _____

2. Which **section** represents the liquid phase? _____

3. Which **section** represents the gas phase? _____

4. Which **letter** represents the triple point? _____

5. Which **letter** represents the critical point? _____

6. What is this substance's normal melting point? _____

7. What is this substance's normal boiling point? _____

8. Above what temperature is it impossible to liquify this substance no matter what the pressure? _____

9. At what temperature and pressure do all three phases coexist? _____

10. Is the density of the solid greater than or less than the density of the liquid?

    _____

11. Would an increase in pressure cause this substance to freeze or melt? _____

Heat and Its Measurement

Heat (or energy) can be measured in units of **calories** or **joules**. When there is a temperature change (ΔT), heat (Q) can be calculated using this formula:

Q = mass × ΔT × specific heat capacity
(ΔT = final temperature – initial temperature)

During a phase change, use this formula:

Q = mass × heat of fusion (or heat of vaporization)

Solve each problem.

1.	How many joules of heat are given off when 5.0 g of water cool from 75°C to 25°C? (Specific heat of water = 4.18 J/g°C)
2.	How many calories are given off by the water in problem 1? (Specific heat of water = 1.0 cal/g°C)
3.	How many joules does it take to melt 35 g of ice at 0°C? (heat of fusion = 333 J/g)
4.	How many calories are given off when 85 g of steam condense to liquid water? (heat of vaporization = 539.4 cal/g)
5.	How many joules of heat are necessary to raise the temperature of 25 g of water from 10°C to 60°C?
6.	How many calories are given off when 50 g of water at 0°C freezes? (heat of fusion = 79.72 cal/g)

Vapor Pressure and Boiling

A liquid will boil when its vapor pressure equals the atmospheric pressure.

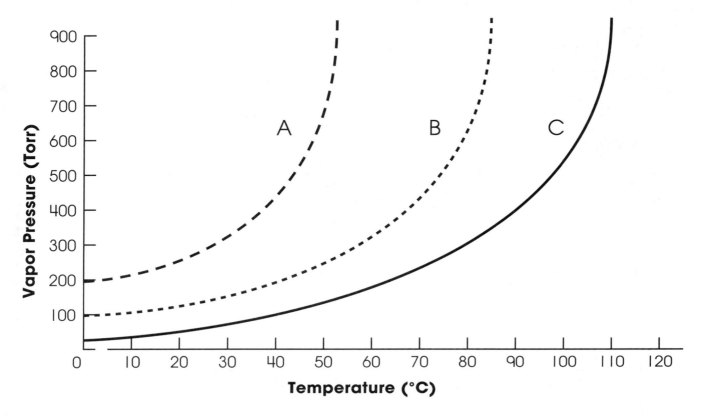

Use the graph to answer each question.

1. At what temperature would Liquid A boil at an atmospheric pressure of 400 Torr? _____

2. Liquid B? _____

3. Liquid C? _____

4. How low must the atmospheric pressure be for Liquid A to boil at 35°C? _____

5. Liquid B? _____

6. Liquid C? _____

7. What is the normal boiling point of Liquid A? _____

8. Liquid B? _____

9. Liquid C? _____

10. Which liquid has the strongest intermolecular forces? _____

Matter—Substances vs. Mixtures

All matter can be classified as either a **substance** (element or compound) or a **mixture** (heterogeneous or homogeneous).

Matter

Substance
can write chemical
formula, homogeneous

Element
one type
of atom

Compound
two or more different
atoms, chemically bonded

Mixture
variable ratio

Homogeneous
solutions

Heterogeneous
colloids and
suspensions

Classify each of the following as a substance or a mixture. If it is a substance, write *element* or *compound* in the substance column. If it is a mixture, write *heterogeneous* or *homogeneous* in the mixture column.

Type of Matter	Substance	Mixture
1. chlorine		
2. water		
3. soil		
4. sugar water		
5. oxygen		
6. carbon dioxide		
7. rocky road ice cream		
8. alcohol		
9. pure air		
10. iron		

Physical vs. Chemical Properties

A **physical property** is observed with the senses and can be determined without destroying the object. Color, shape, mass, length, and odor are all examples of physical properties.

A **chemical property** indicates how a substance reacts with something else. The original substance is fundamentally changed in observing a chemical property. For example, the ability of iron to rust is a chemical property. The iron has reacted with oxygen, and the original iron metal is changed. It now exists as iron oxide, a different substance.

Classify each property as either chemical or physical by putting a check in the appropriate column.

	Physical Property	Chemical Property
1. blue color		
2. density		
3. flammability		
4. solubility		
5. reacts with acid to form H_2		
6. supports combustion		
7. sour taste		
8. melting point		
9. reacts with water to form a gas		
10. reacts with a base to form water		
11. hardness		
12. boiling point		
13. can neutralize a base		
14. luster		
15. odor		

Physical vs. Chemical Changes

In a **physical change**, the original substance still exists; it only changes in form. In a **chemical change**, a new substance is produced. Energy changes always accompany chemical changes.

Classify each as a *physical* or *chemical* change.

1. Sodium hydroxide dissolves in water. _____

2. Hydrochloric acid reacts with potassium hydroxide to produce a salt, water, and heat. _____

3. A pellet of sodium is sliced in two. _____

4. Water is heated and changed to steam. _____

5. Potassium chlorate decomposes to potassium chloride and oxygen gas. _____

6. Iron rusts. _____

7. When placed in H_2O, a sodium pellet catches on fire as hydrogen gas is liberated and sodium hydroxide forms. _____

8. Water evaporates. _____

9. Ice melts. _____

10. Milk sours. _____

11. Sugar dissolves in water. _____

12. Wood rots. _____

13. Pancakes are cooking on a griddle. _____

14. Grass is growing in a lawn. _____

15. A tire is inflated with air. _____

16. Food is digested in the stomach. _____

17. Water is absorbed by a paper towel. _____

Boyle's Law

Boyle's Law states that the volume of a given sample of gas at a constant temperature varies inversely with the pressure. (If one goes up, the other goes down.) Use the formula:

$$P_1 \times V_1 = P_2 \times V_2$$

Solve each problem (assuming constant temperature).

1. A sample of oxygen gas occupies a volume of 250. mL at 740. Torr. What volume will it occupy at 800. Torr pressure?

   _____

2. A sample of carbon dioxide occupies a volume of 3.50 liters at 125 kPa pressure. What pressure would the gas exert if the volume was decreased to 2.00 liters?

   _____

3. A 2.0 liter container of nitrogen has a pressure of 3.2 atm. What volume would be necessary to decrease the pressure to 1.0 atm?

   _____

4. Ammonia gas occupies a volume of 450. mL at a pressure of 720. mmHg. What volume will it occupy at standard pressure?

   _____

5. A 175 mL sample of neon has its pressure changed from 75 kPa to 150 kPa. What is its new volume?

   _____

6. A sample of hydrogen at 1.5 atm has its pressure decreased to 0.50 atm, producing a new volume of 750 mL. What was its original volume?

   _____

7. Chlorine gas occupies a volume of 1.2 liters at 720 Torr. What volume will it occupy at 1 atm pressure?

   _____

8. Fluorine gas exerts a pressure of 900. Torr. When the pressure is changed to 1.50 atm, its volume is 250. mL. What was the original volume?

   _____

Name_____

Charles' Law

Charles' Law states that the volume of a given sample of gas at a constant pressure is directly proportional to the temperature in Kelvin. Use the following formulas:

$$\frac{V_1}{T_1} = \frac{V_2}{T_2} \quad \text{or} \quad V_1 \times T_2 = V_2 \times T_1$$

$$K = °C + 273$$

Solve each problem (assuming constant pressure).

1. A sample of nitrogen occupies a volume of 250 mL at 25°C. What volume will it occupy at 95°C?

    _____

2. Oxygen gas is at a temperature of 40°C when it occupies a volume of 2.3 liters. To what temperature should it be raised to occupy a volume of 6.5 liters?

    _____

3. Hydrogen gas was cooled from $1\bar{5}0$°C to $5\bar{0}$°C. Its new volume is 75 mL. What was its original volume?

    _____

4. Chlorine gas occupies a volume of 25 mL at $3\bar{0}0$ K. What volume will it occupy at 600 K?

    _____

5. A sample of neon gas at $5\bar{0}$°C and a volume of 2.5 liters is cooled to 25°C. What is the new volume?

    _____

6. Fluorine gas at $30\bar{0}$ K occupies a volume of $50\bar{0}$ mL. To what temperature should it be lowered to bring the volume to $30\bar{0}$ mL?

    _____

7. Helium occupies a volume of 3.8 liters at -45°C. What volume will it occupy at 45°C?

    _____

8. A sample of argon gas is cooled and its volume went from $38\bar{0}$ mL to $25\bar{0}$ mL. If its final temperature was -55°C, what was its original temperature?

    _____

Combined Gas Law

In practical terms, it is often difficult to hold any of the variables constant. When there is a change in pressure, volume, and temperature, the combined gas law is used.

$$\frac{P_1 \times V_1}{T_1} = \frac{P_2 \times V_2}{T_2} \quad \text{or} \quad P_1 V_1 T_2 = P_2 V_2 T_1$$

Complete the chart.

	P_1	V_1	T_1	P_2	V_2	T_2
1.	1.5 atm	3.0 L	$2\overline{0}°C$	2.5 atm		$3\overline{0}°C$
2.	720 Torr	256 mL	25°C		250 mL	$5\overline{0}°C$
3.	$6\overline{0}0$ mmHg	2.5 L	22°C	760 mmHg	1.8 L	
4.		750 mL	0.0°C	2.0 atm	500 mL	25°C
5.	95 kPa	4.0 L		101 kPa	6.0 L	471 K or 198°C
6.	650. Torr		100°	900. Torr	225 mL	$15\overline{0}°C$
7.	850 mmHg	1.5 L	15°C		2.5 L	$3\overline{0}°C$
8.	125 kPa	125 mL		$10\overline{0}$ kPa	$10\overline{0}$ mL	75°C

Dalton's Law of Partial Pressures

Dalton's Law says that the sum of the individual pressures of all the gases that make up a mixture is equal to the total pressure, or: $P_T = P_1 + P_2 + P_3 + ...$ The partial pressure of each gas is equal to the mole fraction of each gas times the total pressure.

$$P_T = P_1 + P_2 + P_3 + ... \quad \text{or} \quad \frac{\text{moles gas}_x}{\text{total moles}} \times P_T = P_x$$

Solve each problem.

1. A 250. mL sample of oxygen is collected over water at 25°C and 760.0 Torr. What is the pressure of the dry gas alone? (Vapor pressure of water at 25°C = 23.8 Torr)

2. A 32.0 mL sample of hydrogen is collected over water at 20°C and 750.0 Torr. What is the pressure of the dry gas alone? (Vapor pressure of water at 20°C = 17.5 Torr)

3. A 54.0 mL sample of oxygen is collected over water at 23°C and 770.0 Torr. What is the pressure of the dry gas alone? (Vapor pressure of water at 23°C = 21.1 Torr)

4. A mixture of 2.00 moles of H_2, 3.00 moles of NH_3, 4.00 moles of CO_2, and 5.00 moles of N_2 exerts a total pressure of 800 Torr. What is the partial pressure of each gas?

5. The partial pressure of F_2 is 300 Torr in a mixture of gases where the total pressure is 1.00 atm. If there are 1.5 total moles in the mixture, how many moles of F_2 are present?

Ideal Gas Law

The **ideal gas law** describes the state of an ideal gas. While an ideal gas is hypothetical, the ideal gas law can be used to approximate the behavior of many gases under normal conditions. Use the formula:

$PV = nRT$ where P = pressure in atmospheres

V = volume in liters

R = Universal Gas Constant
    0.0821 L•atm/mol•K

n = number of moles of gas T = Kelvin temperature

Use the ideal gas law to solve each problem.

1. How many moles of oxygen will occupy a volume of 2.5 liters at 1.2 atm and 25°C?

    _____

2. What volume will 2.0 moles of nitrogen occupy at 720 Torr and $2\overline{0}$°C?

    _____

3. What pressure will be exerted by 25 g of CO_2 at a temperature of 25°C and a volume of $50\overline{0}$ mL? _____

4. At what temperature will 5.00 g of Cl_2 exert a pressure of 900. Torr at a volume of $75\overline{0}$ mL? _____

5. What is the density of NH_3 at $80\overline{0}$ Torr and 25°C? _____

6. If the density of a gas is 1.2 g/L at 745. Torr and $2\overline{0}$°C, what is its molecular mass?

    _____

7. How many moles of nitrogen gas will occupy a volume of 347 mL at 6680 Torr and 27°C? _____

8. What volume will 454 grams (1 lb.) of hydrogen occupy at 1.05 atm and 25°C?

    _____

9. Find the number of grams of CO_2 that exert a pressure of 785 Torr at a volume of 32.5 L and a temperature of 32°C. _____

10. An elemental gas has a mass of 10.3 g. If the volume is 58.4 L and the pressure is 758 Torr at a temperature of 2.5°C, what is the gas? _____

Graham's Law of Effusion

Graham's Law states that a gas will effuse at a rate that is inversely proportional to the square root of its molecular mass, *MM*.

$$\frac{rate_1}{rate_2} = \sqrt{\frac{MM_2}{MM_1}}$$

Solve each problem.

1.	Under the same conditions of temperature and pressure, how many times faster will hydrogen effuse compared to carbon dioxide?
2.	If the carbon dioxide in problem 1 takes 32 seconds to effuse, how long will the hydrogen take?
3.	What is the relative rate of effusion of NH_3 compared to helium? Does NH_3 effuse faster or slower than helium?
4.	If the helium in problem 3 takes 20 seconds to effuse, how long will NH_3 take?
5.	An unknown gas effuses 0.25 times as fast as helium. What is the molecular mass of the unkown gas?

Element Symbols

An element symbol can stand for one atom of the element or one mole of atoms of the element. (One mole = 6.02×10^{23} atoms of an element.)

Write the symbol for each element.

1. oxygen _____
2. hydrogen _____
3. chlorine _____
4. mercury _____
5. fluorine _____
6. barium _____
7. helium _____
8. uranium _____
9. radon _____

10. sulfur _____
11. plutonium _____
12. calcium _____
13. radium _____
14. cobalt _____
15. zinc _____
16. arsenic _____
17. lead _____
18. iron _____

Write the name of the element that corresponds with each symbol.

19. Kr _____
20. K _____
21. C _____
22. Ne _____
23. Si _____
24. Zr _____
25. Sn _____
26. Pt _____
27. Na _____
28. Al _____

29. Cu _____
30. Ag _____
31. P _____
32. Mn _____
33. I _____
34. Au _____
35. Mg _____
36. Ni _____
37. Br _____
38. Hg _____

Atomic Structure

An atom is made up of protons and neutrons (both found in the nucleus) and electrons (in the surrounding electron cloud). The **atomic number** is equal to the number of protons. The **mass number** is equal to the number of protons plus neutrons.

In a neutral atom, the number of protons equals the number of electrons. The **charge** on an ion indicates an imbalance between protons and electrons. Too many electrons produce a negative charge. Too few electrons produce a positive charge.

This structure can be written as part of a chemical symbol.

mass number → $^{15}_{7}N^{3+}$ ← charge
atomic number →

7 protons
8 neutrons (15 – 7)
4 electrons

Complete the chart.

Element/ Ion	Atomic Number	Atomic Mass	Mass Number	Protons	Neutrons	Electrons
H						
H⁺						
$^{12}_{6}C$						
$^{7}_{3}Li^{+}$						
$^{35}_{17}Cl^{-}$						
$^{39}_{19}K$						
$^{24}_{12}Mg^{2+}$						
As³⁻						
Ag						
Ag¹⁺						
S²⁻						
U						

Isotopes and Average Atomic Mass

Elements come in a variety of **isotopes**, meaning they are made up of atoms with the same atomic number but different atomic masses. These atoms differ in the number of neutrons.

The **average atomic mass** is the weighted average of all of the isotopes of an element.

Example: A sample of cesium is 75% ^{133}Cs, 20% ^{132}Cs, and 5% ^{134}Cs. What is its average atomic mass?

Answer: $0.75 \times 133 = $ 99.75
$0.20 \times 132 = $ 26.4
$0.05 \times 134 = $ 6.7
Total = 132.85 amu = average atomic mass

Determine the average atomic mass of each mixture of isotopes.

1. 80% ^{127}I, 17% ^{126}I, 3% ^{128}I

2. 50% ^{197}Au, 50% ^{198}Au

3. 15% ^{55}Fe, 85% ^{56}Fe

4. 99% ^{1}H, 0.8% ^{2}H, 0.2% ^{3}H

5. 95% ^{14}N, 3% ^{15}N, 2% ^{16}N

6. 98% ^{12}C, 2% ^{14}C

Electron Configuration (Level One)

Electrons are distributed in the electron cloud into principal energy levels (1, 2, 3, ...), sublevels (*s, p, d, f*), orbitals (*s* has 1, *p* has 3, *d* has 5, *f* has 7), and spin (two electrons allowed per orbital).

Example: Draw the electron configuration of sodium (atomic number 11).

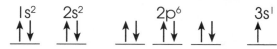

Draw the electron configuration of each atom.

1. Cl
2. N
3. Al
4. O

Electron Configuration (Level Two)

At atomic numbers greater than 18, the sublevels begin to fill out of order. A good approximation of the order of filling can be determined using the diagonal rule.

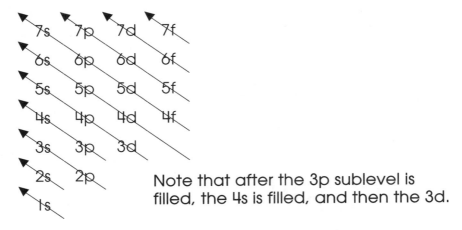

Note that after the 3p sublevel is filled, the 4s is filled, and then the 3d.

Draw the electron configuration of each atom.

1. K
2. V
3. Co
4. Zr

Valence Electrons

The **valence electrons** are the electrons in the outermost principal energy level. They are always s electrons or s and p electrons. Since the total number of electrons possible in s and p sublevels is eight, there can be no more than eight valence electrons.

Example: carbon

Electron configuration is $1s^2$ $\boxed{2s^2 \quad 2p^2}$.

Carbon has 4 valence electrons.

Determine the number of valence electrons in each atom.

1. fluorine _____

2. phosphorus _____

3. calcium _____

4. nitrogen _____

5. iron _____

6. argon _____

7. potassium _____

8. helium _____

9. magnesium _____

10. sulfur _____

11. lithium _____

12. zinc _____

13. carbon _____

14. iodine _____

15. oxygen _____

16. barium _____

17. aluminum _____

18. hydrogen _____

19. xenon _____

20. copper _____

Lewis Dot Diagrams

Lewis dot diagrams are a way to indicate the number of valence electrons around an atom.

Examples: Na•   :C̈l: •N̈:

Draw the Lewis dot diagram of each atom.

1. calcium

2. potassium

3. argon

4. aluminum

5. bromine

6. carbon

7. helium

8. oxygen

9. phosphorus

10. hydrogen

Name_____

Atomic Structure Crossword

Across

1. The smallest particle of an element that can enter into chemical change

4. The number of protons in the nucleus of an atom

6. Cannot be decomposed into simpler substances by ordinary chemical means

7. State in which all electrons are at their lowest possible energy level

8. The positively charged particle found in the nucleus

11. Standard atomic mass unit for carbon

13. Most of the mass of an atom is here.

16. Mass number minus atomic number

18. Electrons in the outermost principal energy level

19. Protons and neutrons are these.

Down

2. Sum of the protons and neutrons in the nucleus of an atom

3. Charged atom or group of atoms

5. Equal to the number of protons in a neutral atom

9. The volume of an atom is determined by the size of its electron.

10. Different forms of the same element

12. State in which electrons have absorbed energy and "jumped" to a higher energy level

14. Atoms with the same atomic number but different atomic masses

15. The nucleus and all electrons in an atom except the valence electrons

17. *s, p, d, f*

Nuclear Decay

Predict the product of each nuclear reaction.

1. $^{42}K \longrightarrow \, _{-1}^{0}e +$ _____

2. $^{239}Pu \longrightarrow \, _{2}^{4}He +$ _____

3. $_{92}^{235}U \longrightarrow$ _____ $+ \, _{90}^{231}Th$

4. $_{1}^{1}H + \, _{1}^{3}H \longrightarrow$ _____

5. $_{3}^{6}Li + \, _{0}^{1}n \longrightarrow \, _{2}^{4}He +$ _____

6. $_{13}^{27}Al + \, _{2}^{4}He \longrightarrow \, _{15}^{30}P +$ _____

7. $_{4}^{9}Be + \, _{1}^{1}H \longrightarrow$ _____ $+ \, _{2}^{4}He$

8. $^{37}K \longrightarrow \, _{+1}^{0}e +$ _____

9. _____ $+ \, _{0}^{1}n \longrightarrow \, _{56}^{142}Ba + \, _{36}^{91}Kr + 3\,_{0}^{1}n$

10. $_{92}^{238}U + \, _{2}^{4}He \longrightarrow$ _____ $+ \, _{0}^{1}n$

Half-Lives of Radioactive Isotopes

Solve each problem.

1. How much of a 100.0 g sample of ^{198}Au is left after 8.10 days if its half-life is 2.70 days?

2. A 50.0 g sample of ^{16}N decays to 12.5 g in 14.4 seconds. What is its half-life?

3. The half-life of ^{42}K is 12.4 hours. How much of a $75\overline{0}$ g sample is left after 62.0 hours?

4. What is the half-life of ^{99}Tc if a $50\overline{0}$ g sample decays to 62.5 g in 639,000 years?

5. The half-life of ^{232}Th is 1.4×10^{10} years. If there are 25.0 g of the sample left after 2.8×10^{10} years, how many grams were in the original sample?

6. There are 5.0 g of ^{131}I left after 40.35 days. How many grams were in the original sample if its half-life is 8.07 days?

Periodic Table Worksheet

Use a copy of the periodic table to answer each question.

1. Where are the most active metals located? _____

2. Where are the most active nonmetals located? _____

3. As you go from left to right across a period, the atomic size (decreases, increases). Why? _____

4. As you travel down a group, the atomic size (decreases, increases). Why?

   _____

5. A negative ion is (larger, smaller) than its parent atom.

6. A positive ion is (larger, smaller) than its parent atom.

7. As you go from left to right across a period, the first ionization energy generally (decreases, increases). Why? _____

8. As you go down a group, the first ionization energy generally (decreases, increases). Why? _____

9. Where is the highest electronegativity found? _____

10. Where is the lowest electronegativity found? _____

11. Elements of Group 1 are called _____.

12. Elements of Group 2 are called _____.

13. Elements of Group 3–12 are called _____

14. As you go from left to right across the periodic table, the elements go from (metals, nonmetals) to (metals, nonmetals).

15. Group 17 elements are called _____

16. The most active element in Group 17 is _____

17. Group 18 elements are called _____

18. What sublevels are filling across the Transition Elements? _____

19. Elements within a group have a similar number of _____.

20. Elements across a series have the same number of _____

21. A colored ion generally indicates a _____.

22. As you go down a group, the elements generally become (more, less) metallic.

23. The majority of elements in the periodic table are (metals, nonmetals).

24. Elements in the periodic table are arranged according to their _____

25. An element with both metallic and nonmetallic properties is called a _____

    _____

Periodic Table Puzzle

	1	2	3	4	5	6	7	8	9	10	11	12	13	14	15	16	17	18
	I																	
		F														G	H	
														B				A
	C							E				J						

			D												

Place the letter of each of the above elements next to its description.

1. An alkali metal _____

2. An alkaline earth metal _____

3. An inactive gas _____

4. An active nonmetal _____

5. A semimetal _____

6. An inner transition element _____

7. Its most common oxidation state is -2. _____

8. A metal with more than one oxidation state _____

9. A metal with an oxidation number of +3 _____

10. Has oxidation numbers of +1 and -1 _____

Periodic Table Puzzle

	1	2	3	4	5	6	7	8	9	10	11	12	13	14	15	16	17	18
	A													E				
																	D	
			C												G			B
								H			J							
		F																

										I				

Place the letter of each of the above elements next to its description. Then, using a copy of the periodic table, write the element's name and symbol.

1. An element in Group 17 _____

2. An alkali metal with an oxidation of +1 _____

3. A metalloid _____

4. Has 80 electrons _____

5. Its electron configuration is 2, 4. _____

6. Has 99 protons _____

7. An element in the fifth period with fewer than 50 protons _____

8. Its electron configuration is 2, 8, 18, 18, 8. _____

9. A transition metal with one electron in its outer shell _____

10. An alkaline earth metal _____

Ionic Bonding

Ionic bonding occurs when a metal transfers one or more electrons to a nonmetal in an effort to attain a stable octet of electrons. For example, the transfer of an electron from sodium to chlorine can be depicted by a Lewis dot diagram.

$$Na\overset{\frown}{\cdot} + \cdot \ddot{\underset{\cdot\cdot}{C}}l\colon \longrightarrow Na^+Cl^-$$

Calcium would need two chlorine atoms to get rid of its two valence electrons.

$$:\ddot{\underset{\cdot\cdot}{C}}l\cdot + \cdot Ca\overset{\frown}{\cdot} + \cdot\ddot{\underset{\cdot\cdot}{C}}l\colon \longrightarrow Ca^{+2}Cl_2$$

Sketch the transfer of electrons in each combination.

1. K + F
2. Mg + I
3. Be + S
4. Na + O
5. Al + Br

Covalent Bonding

Covalent bonding occurs when two or more nonmetals share electrons, attempting to attain a stable octet of electrons at least part of the time.

Example:

H• + ×C̈l× ⟶ (H××C̈l×)   Note that hydrogen is content with 2, not 8, electrons.

Sketch how covalent bonding occurs in each pair of atoms. Atoms may share one, two, or three pairs of electrons.

1. H + H (H_2)	
2. F + F (F_2)	
3. O + O (O_2)	
4. N + N (N_2)	
5. C + O (CO_2)	
6. H + O (H_2O)	

Types of Chemical Bonds

Identify each compound as *ionic* (metal + nonmetal), *covalent* (nonmetal + nonmetal), or *both* (compound containing a polyatomic ion).

1. $CaCl_2$ _____

2. CO_2 _____

3. H_2O _____

4. $BaSO_4$ _____

5. K_2O _____

6. NaF _____

7. Na_2CO_3 _____

8. CH_4 _____

9. SO_3 _____

10. $LiBr$ _____

11. MgO _____

12. NH_4Cl _____

13. HCl _____

14. KI _____

15. $NaOH$ _____

16. NO_2 _____

17. $AlPO_4$ _____

18. $FeCl_3$ _____

19. P_2O_5 _____

20. N_2O_3 _____

Shapes of Molecules

Using VSEPR theory, name and sketch the shape of each molecule.

1. N_2	7. HF
2. H_2O	8. CH_3OH
3. CO_2	9. H_2S
4. NH_3	10. I_2
5. CH_4	11. $CHCl_3$
6. SO_3	12. O_2

Polarity of Molecules

Identify each molecule as *polar* or *nonpolar*.

1. N_2	7. HF
2. H_2O	8. CH_3OH
3. CO_2	9. H_2S
4. NH_3	10. I_2
5. CH_4	11. $CHCl_3$
6. SO_3	12. O_2

Name_____

Chemical Bonding Crossword

Across

1. Ammonia is polar because its shape is _____.

3. Word used to describe a molecule with an unequal charge distribution

6. Type of bond formed between an active metal and a nonmetal

8. The simultaneous attraction of electrons for the nucleii of two or more atoms is a chemical _____.

11. Type of covalent bond in which one atom donates both electrons

14. Bonding that is responsible for the relatively high boiling point of water

15. Type of covalent bond found in diatomic molecules

16. Carbon dioxide is nonpolar because it is _____.

17. Particles formed from covalent bonding

Down

1. Compounds with both ionic and covalent bonds contain this type of ion.

2. Type of bond found in aluminum foil

4. The formulas of ionic compounds must be expressed as _____ formulas.

5. The shape of a water molecule

7. Type of bond found between nonmetals

9. Type of covalent bonding that is found in a diamond

10. Type of covalent bond found between atoms of different electronegativity values

12. Force of attraction between nonpolar molecules

13. Element with the highest electronegativity value

Writing Formulas (Crisscross Method)

Write the formula of the compound produced from the listed ions.

	Cl^-	CO_3^{-2}	OH^-	SO_4^{-2}	PO_4^{-3}	NO_3^-
Na^+						
NH_4^+						
K^+						
Ca^{2+}						
Mg^{2+}						
Zn^{2+}						
Fe^{3+}						
Al^{3+}						
Co^{3+}						
Fe^{2+}						
H^+						

Naming Ionic Compounds

Name each compound using the Stock Naming System.

1. $CaCO_3$ _____

2. KCl _____

3. $FeSO_4$ _____

4. $LiBr$ _____

5. $MgCl_2$ _____

6. $FeCl_3$ _____

7. $Zn_3(PO_4)_2$ _____

8. NH_4NO_3 _____

9. $Al(OH)_3$ _____

10. $CuC_2H_3O_2$ _____

11. $PbSO_3$ _____

12. $NaClO_3$ _____

13. CaC_2O_4 _____

14. Fe_2O_3 _____

15. $(NH_4)_3PO_4$ _____

16. $NaHSO_4$ _____

17. Hg_2Cl_2 _____

18. $Mg(NO_2)_2$ _____

19. $CuSO_4$ _____

20. $NaHCO_3$ _____

21. $NiBr_3$ _____

22. $Be(NO_3)_2$ _____

23. $ZnSO_4$ _____

24. $AuCl_3$ _____

25. $KMnO_4$ _____

Name____________________________

Naming Molecular Compounds

Name each covalent compound.

1. CO_2 ______________________
2. CO ______________________
3. SO_2 ______________________
4. SO_3 ______________________
5. N_2O ______________________
6. NO ______________________
7. N_2O_3 ______________________
8. NO_2 ______________________
9. N_2O_4 ______________________
10. N_2O_5 ______________________
11. PCl_3 ______________________
12. PCl_5 ______________________
13. NH_3 ______________________
14. SCl_6 ______________________
15. P_2O_5 ______________________
16. CCl_4 ______________________
17. SiO_2 ______________________
18. CS_2 ______________________
19. OF_2 ______________________
20. PBr_3 ______________________

Naming Acids

Name each acid.

1. HNO_3 _____

2. HCl _____

3. H_2SO_4 _____

4. H_2SO_3 _____

5. $HC_2H_3O_2$ _____

6. HBr _____

7. HNO_2 _____

8. H_3PO_4 _____

9. H_2S _____

10. H_2CO_3 _____

Write the formula of each acid.

11. sulfuric acid _____

12. nitric acid _____

13. hydrochloric acid _____

14. acetic acid _____

15. hydrofluoric acid _____

16. phosphorous acid _____

17. carbonic acid _____

18. nitrous acid _____

19. phosphoric acid _____

20. hydrosulfuric acid _____

Writing Formulas from Names

Write the formula of each compound.

1. ammonium phosphate _____

2. iron(II) oxide _____

3. iron(III) oxide _____

4. carbon monoxide _____

5. calcium chloride _____

6. potassium nitrate _____

7. magnesium hydroxide _____

8. aluminum sulfate _____

9. copper(II) sulfate _____

10. lead(IV) chromate _____

11. diphosphorus pentoxide _____

12. potassium permanganate _____

13. sodium hydrogen carbonate _____

14. zinc nitrate _____

15. aluminum sulfite _____

Name_____

Gram Formula Mass

Determine the gram formula mass (the mass of one mole) of each compound.

1. $KMnO_4$ _____

2. KCl _____

3. Na_2SO_4 _____

4. $Ca(NO_3)_2$ _____

5. $Al_2(SO_4)_3$ _____

6. $(NH_4)_3PO_4$ _____

7. $CuSO_4 \cdot 5H_2O$ _____

8. $Mg_3(PO_4)_2$ _____

9. $Zn(C_2H_3O_2)_2 \cdot 2H_2O$ _____

10. $Zn_3(PO_4)_2 \cdot 4H_2O$ _____

11. H_2CO_3 _____

12. $Hg_2Cr_2O_7$ _____

13. $Ba(ClO_3)_2$ _____

14. $Fe_2(SO_3)_3$ _____

15. $NH_4C_2H_3O_2$ _____

Moles and Mass

Determine the number of moles in each quantity.

1.   25 g of NaCl
2.   125 g of H_2SO_4
3.   100. g of $KMnO_4$
4.   74 g of KCl
5.   35 g of $CuSO_4 \cdot 5H_2O$

Determine the number of grams in each quantity.

6.   2.5 mol of NaCl
7.   0.50 mol of H_2SO_4
8.   1.70 mol of $KMnO_4$
9.   0.25 mol of KCl
10.   3.2 mol of $CuSO_4 \cdot 5H_2O$

The Mole and Volume

For gases at STP (273 K and 1 atm pressure), one mole occupies a volume of 22.4 L. Identify the volume each quantity of gas will occupy at STP.

1.	1.00 mole of H_2
2.	3.20 moles of O_2
3.	0.750 mole of N_2
4.	1.75 moles of CO_2
5.	0.50 mole of NH_3
6.	5.0 g of H2
7.	100. g of O_2
8.	28.0 g of N_2
9.	60. g of CO_2
10.	10. g of NH_3

The Mole and Avogadro's Number

One mole of a substance contains Avogadro's number (6.02×10^{23}) of molecules.

Identify how many molecules are in each quantity.

1.	2.0 mol
2.	1.5 mol
3.	0.75 mol
4.	15 mol
5.	0.35 mol

Identify how many moles are in the number of molecules listed.

6.	6.02×10^{23}
7.	1.204×10^{24}
8.	1.5×10^{20}
9.	3.4×10^{26}
10.	7.5×10^{19}

Mixed Mole Problems

Solve each problem.

1. How many grams are there in 1.5×10^{25} molecules of CO_2?

2. What volume would the CO_2 in problem 1 occupy at STP?

3. A sample of NH_3 gas occupies 75.0 liters at STP. How many molecules is this?

4. What is the mass of the sample of NH_3 in problem 3?

5. How many atoms are there in 1.3×10^{22} molecules of NO_2?

6. A 5.0 g sample of O_2 is in a container at STP. What volume is the container?

7. How many molecules of O_2 are in the container in problem 6? How many atoms of oxygen?

Percentage Composition

Determine the percentage composition of each compound.

1. $KMnO_4$

   K = _____

   Mn = _____

   O = _____

2. HCl

   H = _____

   Cl = _____

3. $Mg(NO_3)_2$

   Mg = _____

   N = _____

   O = _____

4. $(NH_4)PO_4$

   N = _____

   H = _____

   P = _____

   O = _____

5. $Al_2(SO_4)_3$

   Al = _____

   S = _____

   O = _____

Solve each problem.

6. How many grams of oxygen can be produced from the decomposition of 100. g of $KClO_3$? _____

7. How much iron can be recovered from 25.0 g of Fe_2O_3? _____

8. How much silver can be produced from 125 g of Ag_2S? _____

Determining Empirical Formulas

Identify the empirical formula (lowest whole number ratio) of each compound.

1.	75% carbon, 25% hydrogen
2.	52.7% potassium, 47.3% chlorine
3.	22.1% aluminum, 25.4% phosphorus, 52.5% oxygen
4.	13% magnesium, 87% bromine
5.	32.4% sodium, 22.5% sulfur, 45.1% oxygen
6.	25.3% copper, 12.9% sulfur, 25.7% oxygen, 36.1% water

Determining Molecular Formulas
(True Formulas)

Solve each problem.

1.	The empirical formula of a compound is NO_2. Its molecular mass is 92 g/mol. What is its molecular formula?
2.	The empirical formula of a compound is CH_2. Its molecular mass is 70 g/mol. What is its molecular formula?
3.	A compound is found to be 40.0% carbon, 6.7% hydrogen and 53.5% oxygen. Its molecular mass is 60. g/mol. What is its molecular formula?
4.	A compound is 64.9% carbon, 13.5% hydrogen, and 21.6% oxygen. Its molecular mass is 74 g/mol. What is its molecular formula?
5.	A compound is 54.5% carbon, 9.1% hydrogen, and 36.4% oxygen. Its molecular mass is 88 g/mol. What is its molecular formula?

Composition of Hydrates

A **hydrate** is an ionic compound with water molecules loosely bonded to its crystal structure. The water is in a specific ratio to each formula unit of the salt. For example, the formula $CuSO_4 \cdot 5H_2O$ indicates that there are five water molecules for every one formula unit of $CuSO_4$.

Solve each problem.

1.	What percentage of water is found in $CuSO_4 \cdot 5H_2O$?
2.	What percentage of water is found in $Na_2S \cdot 9H_2O$?
3.	A 5.0 g sample of a hydrate of $BaCl_2$ was heated, and only 4.3 g of the anhydrous salt remained. What percentage of water was in the hydrate?
4.	A 2.5 g sample of a hydrate of $Ca(NO_3)_2$ was heated, and only 1.7 g of the anhydrous salt remained. What percentage of water was in the hydrate?
5.	A 3.0 g sample of $Na_2CO_3 \cdot H_2O$ is heated to constant mass. How much anhydrous salt remains?
6.	A 5.0 g sample of $Cu(NO_3)_2 \cdot nH_2O$ is heated to constant mass. How much anhydrous salt remains?

Balancing Chemical Equations

Rewrite and balance each equation.

1. $N_2 + H_2 \rightarrow NH_3$ _____

2. $KClO_3 \rightarrow KCl + O_2$ _____

3. $NaCl + F_2 \rightarrow NaF + Cl_2$ _____

4. $H_2 + O_2 \rightarrow H_2O$ _____

5. $AgNO_3 + MgCl_2 \rightarrow AgCl + Mg(NO_3)_2$ _____

6. $AlBr_3 + K_2SO \rightarrow KBr + Al_2(SO_4)_3$ _____

7. $CH_4 + O_2 \rightarrow CO_2 + H_2O$ _____

8. $C_3H_8 + O_2 \rightarrow CO_2 + H_2O$ _____

9. $C_8H_{18} + O_2 \rightarrow CO_2 + H_2O$ _____

10. $FeCl_3 + NaOH \rightarrow Fe(OH)_3 + NaCl$ _____

11. $P + O_2 \rightarrow P_2O_5$ _____

12. $Na + H_2O \rightarrow NaOH + H_2$ _____

13. $Ag_2O \rightarrow Ag + O_2$ _____

14. $S_8 + O_2 \rightarrow SO_3$ _____

15. $CO_2 + H_2O \rightarrow C_6H_{12}O_6 + O_2$ _____

16. $K + MgBr_2 \rightarrow KBr + Mg$ _____

17. $HCl + CaCO_3 \rightarrow CaCl_2 + H_2O + CO_2$ _____

Word Equations

Rewrite each word equation as a chemical equation. Then, balance the equation.

1. zinc + lead(II) nitrate yield zinc nitrate + lead

2. aluminum bromide + chlorine yield aluminum chloride + bromine

3. sodium phosphate + calcium chloride yield calcium + sodium chloride

4. potassium chlorate, when heated, yields potassium chloride + oxygen gas

5. aluminum + hydrochloric acid yield aluminum chloride + hydrogen gas

6. calcium hydroxide + phosphoric acid yield calcium phosphate + water

7. copper + sulfuric acid yield copper(II) sulfate + water + sulfur dioxide

8. hydrogen + nitrogen monoxide yield water + nitrogen

Classification of Chemical Reactions

Identify each reaction as *synthesis, decomposition, cationic* or *anionic single replacement*, or *double replacement*.

1. $2H_2 + O_2 \rightarrow 2H_2O$

2. $2H_2O \rightarrow 2H_2 + O_2$

3. $Zn + H_2SO_4 \rightarrow ZnSO_4 + H_2$

4. $2CO + O_2 \rightarrow 2CO_2$

5. $2HgO \rightarrow 2Hg + O_2$

6. $2KBr + Cl_2 \rightarrow 2KCl + Br_2$

7. $CaO + H_2O \rightarrow Ca(OH)_2$

8. $AgNO + NaCl \rightarrow AgCl + NaNO_3$

9. $2H_2O_2 \rightarrow 2H_2O + O_2$

10. $Ca(OH)_2 + H_2SO_4 \rightarrow CaSO_4 + 2H_2O$

Predicting Products of Chemical Reactions

Predict the product in each reaction. Then, write the balanced equation and classify the reaction.

1.   magnesium bromide + chlorine
2.   aluminum + iron(III) oxide
3.   silver nitrate + zinc chloride
4.   hydrogen peroxide (catalyzed by manganese dioxide)
5.   zinc + hydrochloric acid
6.   sulfuric acid + sodium hydroxide
7.   sodium + hydrogen
8.   acetic acid + copper

Stoichiometry: Mole-Mole Problems

Solve each problem.

1. $N_2 + 3H_2 \longrightarrow 2NH_3$

   How many moles of hydrogen are needed to completely react with two moles of nitrogen?

2. $2KClO_3 \longrightarrow 2KCl + 3O_2$

   How many moles of oxygen are produced by the decomposition of six moles of potassium chlorate?

3. $Zn + 2HCl \longrightarrow ZnCl_2 + H_2$

   How many moles of hydrogen are produced from the reaction of three moles of zinc with an excess of hydrochloric acid?

4. $C_3H_8 + 5O_2 \longrightarrow 3CO_2 + 4H_2O$

   How many moles of oxygen are necessary to react completely with four moles of propane (C_3H_8)?

5. $K_3PO_4 + Al(NO_3)_3 \longrightarrow 3KNO_3 + AlPO_4$

   How many moles of potassium nitrate (KNO_3) are produced when six moles of potassium phosphate (KPO_4) react with two moles of aluminum nitrate ($Al(NO_3)_3$)?

Stoichiometry: Volume-Volume Problems

Solve each problem.

1. $N_2 + 3H_2 \rightarrow 2NH_3$
   What volume of hydrogen is necessary to react with five liters of nitrogen to produce ammonia? (Assume constant temperature and pressure.)

2. What volume of ammonia is produced in the reaction in problem 1?

3. $C_3H_8 + 5O_2 \rightarrow 3CO_2 + 4H_2O$
   If 20 liters of oxygen are consumed in the above reaction, how many liters of carbon dioxide are produced?

4. $2H_2O \rightarrow 2H + O_2$
   If 30 mL of hydrogen are produced in the above reaction, how many milliliters of oxygen are produced?

5. $2CO + O_2 \rightarrow 2CO_2$
   How many liters of carbon dioxide are produced if 75 liters of carbon monoxide are burned in oxygen? How many liters of oxygen are necessary?

Stoichiometry: Mass-Mass Problems

Solve each problem.

1. $2KClO_3 \longrightarrow 2KCl + 3O_2$

   How many grams of potassium chloride are produced if 25 g of potassium chlorate decompose?

2. $N_2 + 3H_2 \longrightarrow 2NH_3$

   How many grams of hydrogen are necessary to react completely with 50.0 g of nitrogen in the above reaction?

3. How many grams of ammonia are produced in the reaction in problem 2?

4. $2AgNO_3 + BaCl \longrightarrow 2AgCl + Ba(NO_3)$

   How many grams of silver chloride are produced from 5.0 g of silver nitrate reacting with an excess of barium chloride?

5. How much barium chloride is necessary to react with the silver nitrate in problem 4?

Stoichiometry: Mixed Problems

Solve each problem.

1. $N_2 + 3H_2 \rightarrow 2NH_3$
   What volume of NH_3 at STP is produced if 25.0 g of N_2 is reacted with an excess of H_2?

2. $2KClO_3 \rightarrow 2KCl + 3O_2$
   If 5.0 g of $KClO_3$ are decomposed, what volume of O_2 is produced at STP?

3. How many grams of KCl are produced in problem 2?

4. $Zn + 2HCl \rightarrow ZnCl_2 + H_2$
   What volume of hydrogen at STP is produced when 2.5 g of zinc react with an excess of hydrochloric acid?

5. $H_2SO_4 + 2NaOH \rightarrow H_2O + Na_2SO_4$
   How many molecules of water are produced if 2.0 g of sodium sulfate are produced in the above reaction?

6. $2AlCl_3 \rightarrow 2Al + 3Cl_2$
   If 10.0 g of aluminum chloride are decomposed, how many molecules of Cl_2 are produced?

Stoichiometry: Limiting Reagent

Solve each problem.

1. $N_2 + 3H_2 \rightarrow 2NH_3$

   How many grams of NH_3 can be produced from the reaction of 28 g of N_2 and 25 g of H_2?

2. How much of the excess reagent in problem 1 is left over?

3. $Mg + 2HCl \rightarrow MgCl_2 + H_2$

   What volume of hydrogen at STP is produced from the reaction of 50.0 g of Mg and the equivalent of 75 g of HCl?

4. How much of the excess reagent in problem 3 is left over?

5. $3AgNO_3 + Na_3PO_4 \rightarrow Ag_3PO_4 + 3NaNO_3$

   Silver nitrate and sodium phosphate are reacted in equal amounts of 200. g each. How many grams of silver phosphate are produced?

6. How much of the excess reagent in problem 5 is left over?

Solubility Curves

Answer each question based on the solubility curve shown.

1. Which salt is least soluble in water at 20°C? _____

2. How many grams of potassium chloride can be dissolved in 200 g of water at 80°C? _____

3. At 40°C, how much potassium nitrate can be dissolved in 300 g of water?

   _____

4. Which salt shows the least change in solubility from 0°C to 100°C?

   _____

5. At 30°C, 85 g of sodium nitrate are dissolved in 100 g of water. Is this solution *saturated, unsaturated,* or *supersaturated*? _____

6. A saturated solution of potassium chlorate is formed from 100 g of water. If the saturated solution is cooled from 80°C to 50°C, how many grams of precipitate are formed?

   _____

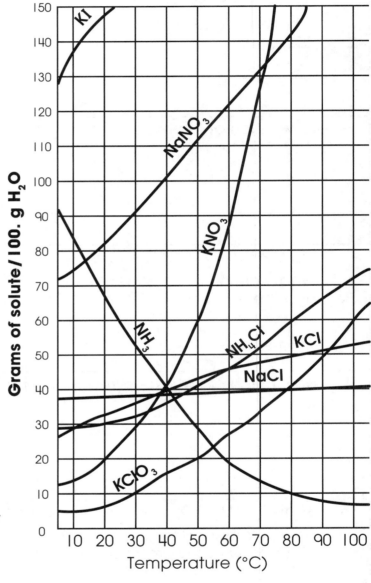

7. What compound shows a decrease in solubility from 0°C to 100°C? _____

8. Which salt is most soluble at 10°C? _____

9. Which salt is least soluble at 50°C? _____

10. Which salt is least soluble at 90°C? _____

Molarity (M)

$$\text{Molarity} = \frac{\text{moles of solute}}{\text{liter of solution}}$$

Solve each problem.

1. What is the molarity of a solution in which 58 g of NaCl are dissolved in 1.0 L of solution?
2. What is the molarity of a solution in which 10.0 g of $AgNO_3$ are dissolved in 500. mL of solution?
3. How many grams of KNO_3 should be used to prepare 2.00 L of a 0.500 M solution?
4. To what volume should 5.0 g of KCl be diluted in order to prepare a 0.25 M solution?
5. How many grams of $CuSO_4 \cdot 5H_2O$ are needed to prepare 100. mL of a 0.10 M solution?

Molarity by Dilution

Acids are usually acquired from chemical supply houses in concentrated form. These acids are diluted to the desired concentration by adding water. Since moles of acid before dilution equal moles of acid after dilution, and moles of acid = $M \times V$, then $M_1 \times V_1 = M_2 \times V_2$.

Solve each problem.

1.	How much concentrated 18 M sulfuric acid is needed to prepare 250 mL of a 6.0 M solution?
2.	How much concentrated 12 M hydrochloric acid is needed to prepare $1\overline{0}0$ mL of a 2.0 M solution?
3.	To what volume should 25 mL of 15 M nitric acid be diluted to prepare a 3.0 M solution?
4.	How much water should be added to 50. mL of 12 M hydrochloric acid to produce a 4.0 M solution?
5.	How much water should be added to 100. mL of 18 M sulfuric acid to prepare a 1.5 M solution?

Molality (m)

$$\text{Molality} = \frac{\text{moles of solute}}{\text{kg of solvent}}$$

Solve each problem.

1. What is the molality of a solution in which 3.0 moles of NaCl are dissolved in 1.5 kg of water?
2. What is the molality of a solution in which 25 g of NaCl are dissolved in 2.0 kg of water?
3. What is the molality of a solution in which 15 g of I_2 are dissolved in 500. g of alcohol?
4. How many grams of I_2 should be added to 750 g of CCl_4 to prepare a 0.020 m solution?
5. How much water should be added to 5.00 g of KCl to prepare a 0.500 m solution?

Normality (N)

normality = molarity × total positive oxidation number of solute

Example: What is the normality of 3.0 M of H_2SO_4?

Since the total positive oxidation number of H_2SO_4 is +2 (2 H^+), N = 6.0.

Solve each problem.

1. What is the normality of a 2.0 M NaOH solution?
2. What is the normality of a 2.0 M H_3PO_4 solution?
3. A solution of H_2SO_4 is 3.0 N. What is its molarity?
4. What is the normality of a solution in which 2.0 g of $Ca(OH)_2$ is dissolved in 1.0 L of solution?
5. How much $AlCl_3$ should be dissolved in 2.00 L of solution to produce a 0.150 N solution?

Electrolytes

Electrolytes are substances that break up (dissociate or ionize) in water to produce ions. These ions are capable of conducting an electric current.

Generally, electrolytes consist of acids, bases, and salts (ionic compounds). Nonelectrolytes are usually covalent compounds, with the exception of acids.

Check the appropriate column to classify each compound as either an electrolyte or a nonelectrolyte.

Compound	Electrolyte	Nonelectrolyte
1. NaCl		
2. CH_3OH (methyl alcohol)		
3. $C_3H_5(OH)_3$ (glycerol)		
4. HCl		
5. $C_6H_{12}O_6$ (sugar)		
6. NaOH		
7. C_2H_5OH (ethyl alcohol)		
8. CH_3COOH (acetic acid)		
9. NH_4OH ($NH_3 + H_2O$)		
10. H_2SO_4		

Effect of a Solute on Freezing and Boiling Points

Use the following formulas to calculate changes in freezing and boiling points due to the presence of a nonvolatile solute. The freezing point is always lowered; the boiling point is always raised.

$$\Delta T_F = M \times d.f. \times k_F \qquad\qquad k_B H_2O = 0.52°C/M$$
$$\Delta T_B = M \times d.f. \times k_B \qquad\qquad k_F H_2O = 1.86°C/M$$

M = molality of solution

k_F and k_B = constants for particular solvent

$d.f.$ = dissociation factor (how many particles the solute breaks up into; for a nonelectrolyte, $d.f.$ = 1)

(The theoretical dissociation factor is always greater than observed effect.)

Solve each problem.

1.   What is the new boiling point if 25 g of NaCl are dissolved in 1.0 kg of water?
2.   What is the freezing point of the solution in problem 1?
3.   What are the new freezing and boiling points of water if 50. g of ethylene glycol (molecular mass = 62 g/M) are added to 50. g of water?
4.   When 5.0 g of a nonelectrolyte are added to 25 g of water, the new freezing point is -2.5°C. What is the molecular mass of the unknown compound?

Solubility (Polar vs. Nonpolar)

Generally, "like dissolves like." Polar molecules dissolve other polar molecules and ionic compounds. Nonpolar molecules dissolve other nonpolar molecules. Alcohols, which have characteristics of both, tend to dissolve in both types of solvents but will not dissolve ionic solids.

Check the appropriate columns to indicate whether the solute is soluble in a polar or nonpolar solvent.

Solutes	Solvents		
	water	CCl_4	alcohol
1. NaCl			
2. I_2			
3. ethanol			
4. benzene			
5. Br_2			
6. KNO_3			
7. toluene			
8. $Ca(OH)_2$			

Solutions Crossword

Across

2. Solution containing the maximum amount of solute possible at that temperature

4. Two liquids which can mix are said to be _____.

6. The presence of a nonvolatile solute will _____ the boiling point of a solvent.

9. A homogeneous mixture

10. Substance present in larger amounts in a mixture

13. Moles of a solute per kilogram of solvent

14. Solution containing a relatively large amount of solvent

15. The solubility of gases _____ as temperature increases.

17. State in which the rate of dissolving is equal to the rate of precipitation

18. The presence of a nonvolatile solute will _____ the freezing point of a solvent.

19. These substances dissociate or ionize in water and are then able to conduct an electric current.

Down

1. Properties that depend on the number of particles in a solution

3. Solution in which more solute can be dissolved

5. Solution containing a relatively large amount of dissolved solute

7. Substance present in a smaller amount in a mixture

8. The solubility of most solids _____ as temperature increases.

11. Maximum amount of solute that can dissolve in a stated amount of solute at a given temperature

12. Moles of solute per liter of solution

16. Solutions in which water is the solvent are called _____.

Potential Energy Diagram

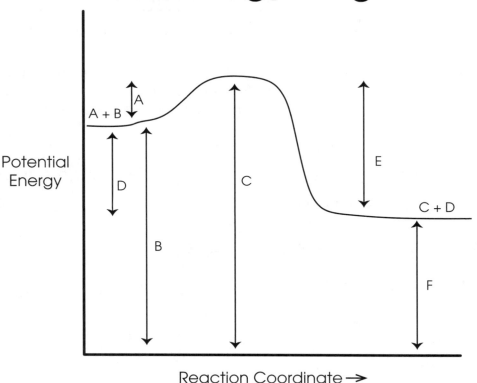

$$A + B \longleftrightarrow C + D + energy$$

Answer each question using the graph shown.

1. Is the above reaction endothermic or exothermic? _____

2. Which letter represents the potential energy of the reactants? _____

3. Which letter represents the potential energy of the products? _____

4. Which letter represents the heat of reaction (ΔH)? _____

5. Which letter represents the activation energy of the forward reaction? _____

6. Which letter represents the activation energy of the reverse reaction? _____

7. Which letter represents the potential energy of the activated complex? _____

8. Is the reverse reaction endothermic or exothermic? _____

9. If a catalyst were added, what letter(s) would change? _____

Entropy

Entropy is the degree of randomness in a substance. The symbol for change in entropy is ΔS.

Solids are very ordered and have low entropy. Liquids and aqueous ions have more entropy because they move about more freely. Gases have an even larger amount of entropy. According to the Second Law of Thermodynamics, nature is always proceeding to a state of higher entropy.

Determine whether each reaction shows an *increase* or *decrease* in entropy.

1. $2KClO_3(s) \rightarrow 2KCl(s) + 3O_2(g)$ _____

2. $H_2O(l) \rightarrow H_2O(s)$ _____

3. $N_2(g) + 3H_2(g) \rightarrow 2NH_3(g)$ _____

4. $NaCl(s) \rightarrow Na^+(aq) + Cl^-(aq)$ _____

5. $KCl(s) \rightarrow KCl(l)$ _____

6. $CO_2(s) \rightarrow CO_2(g)$ _____

7. $H^+(aq) + C_2H_3O_2^-(aq) \rightarrow HC_2H_3O_3(l)$ _____

8. $C(s) + O_2(g) \rightarrow CO_2(g)$ _____

9. $H_2(g) + Cl_2(g) \rightarrow 2HCl(g)$ _____

10. $Ag^+(aq) + Cl^-(aq) \rightarrow AgCl(s)$ _____

11. $2N_2O_5(g) \rightarrow 4NO_2(g) + O_2(g)$ _____

12. $2Al(s) + 3I_2(s) \rightarrow 2AlI_3(s)$ _____

13. $H^+(aq) + OH^-(aq) \rightarrow H_2O(l)$ _____

14. $2NO(g) \rightarrow N_2(g) + O_2(g)$ _____

15. $H_2O(g) \rightarrow H_2O(l)$ _____

Gibbs Free Energy

For a reaction to be spontaneous, the sign of ΔG (Gibbs free energy) must be negative. The mathematical formula for this value is:

$$\Delta G = \Delta H - T\Delta S$$

where ΔH = change in enthalpy or heat of reaction

T = temperature in Kelvin

ΔS = change in entropy or randomness

Complete the table for the sign of ΔG: +, −, or *undetermined*. When conditions allow for an undetermined sign of ΔG, temperature will decide spontaneity.

ΔH	ΔS	ΔG
−	+	
+	−	
−	−	
+	+	

Answer each question.

1. The conditions in which ΔG is always negative are when ΔH is _____ and ΔS is _____.

2. The conditions in which ΔG is always positive are when ΔH is _____ and ΔS is _____.

3. When the situation is indeterminate, a low temperature favors the (entropy, enthalpy) factor and a high temperature favors the (entropy, enthalpy) factor.

Answer problems 4–6 with *always, sometimes,* or *never*.

4. The reaction: $Na(OH)_5 \rightarrow Na^+(aq) + OH^-(aq) + $ energy will _____ be spontaneous.

5. The reaction: energy $+ 2H_2(g) + O_2(g) + 2H_2O(l)$ will _____ be spontaneous.

6. The reaction: energy $+ H_2O(s) \rightarrow H_2O(l)$ will _____ be spontaneous.

7. What is the value of ΔG if $\Delta H = 32.0$ kJ, $\Delta S = +25.0$ kJ/K and T = 293 K? _____

8. Is the reaction in problem 7 spontaneous? _____

9. What is the value of ΔG if $\Delta H = +12.0$ kJ, $\Delta S = 5.00$ kJ/K and T = 290. K? _____

10. Is the reaction in problem 9 spontaneous? _____

Equilibrium Constant (K)

Write the expression for the equilibrium constant (K) for each reaction.

1. $N_2(g) + 3H_2(g) \longleftrightarrow 2NH_3(g)$
2. $2KClO_3(s) \longleftrightarrow 2KCl(s) + 3O_2(g)$
3. $H_2O(l) \longleftrightarrow H^+(aq) + OH^-(aq)$
4. $2CO(g) + O_2(g) \longleftrightarrow 2CO_2(g)$
5. $Li_2CO_3(s) \rightarrow 2Li^+(aq) + CO_3^{2-}(aq)$

Calculations Using the Equilibrium Constant

Using the equilibrium constant expressions you determined on page 80, calculate the value of K when:

1.	$(NH_3) = 0.0100$ M, $(N_2) = 0.0200$ M, $(H_2) = 0.0200$ M
2.	$(O_2) = 0.0500$ M
3.	$(H^+) = 1 \times 10^{-8}$ M, $(OH^-) = 1 \times 10^{-6}$ M
4.	$(CO) = 2.0$ M, $(O_2) = 1.5$ M, $(CO_2) = 3.0$ M
5.	$(Li^+) = 0.2$ M, $(CO_3^{-2}) = 0.1$ M

Le Chatelier's Principle

Le Chatelier's principle states that when a system at equilibrium is subjected to a stress, the system will shift its equilibrium point in order to relieve the stress.

Complete the chart by writing *left, right,* or *none* for equilibrium shift. Then, write *decreases, increases,* or *remains the same* for the concentrations of reactants and products, and for the value of K. The first one has been done for you.

$$N_2(g) + 3H_2(g) \longleftrightarrow 2NH_3(g) + 22.0 \text{ kcal}$$

Stress	Equilibrium Shift	(N_2)	(H_2)	(NH_3)	K
1. Add N_2	right	———	decreases	increases	remains the same
2. Add H_2			———		
3. Add NH_3				———	
4. Remove N_2		———			
5. Remove H_2			———		
6. Remove NH_3				———	
7. Increase temperature					
8. Decrease temperature					
9. Increase pressure					
10. Decrease pressure					

Le Chatelier's Principle (Cont.)

$$12.6 \text{ kcal} + H_2(g) + I_2(g) \leftrightarrow 2HI(g)$$

Stress	Equilibrium Shift	(H_2)	(I_2)	(HI)	K
11. Add H_2	right	———	decreases	increases	remains the same
12. Add I_2			———		
13. Add HI				———	
14. Remove H_2		———			
15. Remove I_2			———		
16. Remove HI				———	
17. Increase temperature					
18. Decrease temperature					
19. Increase pressure					
20. Decrease pressure					

$$NaOH(s) \leftrightarrow Na^+(aq) + OH^-(aq) + 10.6 \text{ kcal}$$

Stress	Equilibrium Shift	Amount NaOH(s)	(Na^+)	(OH^-)	K
21. Add NaOH(s)		———			
22. Add NaCl (Adds Na^+)			———		
23. Add KOH (Adds OH^-)				———	
24. Add H^+ (Removes OH^-)				———	
25. Increase temperature					
26. Decrease temperature					
27. Increase pressure					
28. Decrease pressure					

Bronsted-Lowry Acids and Bases

According to **Bronsted-Lowry theory**, an acid is a proton (H^+) donor and a base is a proton acceptor.

Example: $HCl + OH^- \rightarrow Cl^- + H_2O$

The HCl acts as an acid, and the OH^- acts as a base. This reaction is reversible in that the H_2O can give back the proton to the Cl^-.

Label the Bronsted-Lowry acids and bases in each reaction and show the direction of proton transfer.

Example: $\underset{\text{acid}}{H_2O} + \underset{\text{base}}{Cl^-} \longleftrightarrow \underset{\text{base}}{OH^-} + \underset{\text{acid}}{HCl}$

1. $H_2O + H_2O \longleftrightarrow H_3O^+ + OH^-$

4. $OH^- + H_3O+ \longleftrightarrow H_2O + H_2O$

2. $H_2SO_4 + OH^- \longleftrightarrow HSO_4^- + H_2O$

5. $NH_3 + H_2O \longleftrightarrow NH_4^+ + OH^-$

3. $HSO_4^- + H_2O \longleftrightarrow SO_4^{2-} + H_3O^+$

Conjugate Acid-Base Pairs

In the exercise on page 84, it was shown that after an acid has given up its proton, it is capable of getting the proton back and acting as a base. **Conjugate base** is what is left after an acid gives up a proton. The stronger the acid, the weaker the conjugate base. The weaker the acid, the stronger the conjugate base.

Complete the chart.

Conjugate Pairs

	Acid	Base	Equation
1.	H_2SO_4	HSO_4^-	$H_2SO_4 \longleftrightarrow H^+ + HSO_4^-$
2.	H_3PO_4		
3.		F^-	
4.		NO_3^-	
5.	$H_2PO_4^-$		
6.	H_2O		
7.		SO_4^{2-}	
8.	HPO_4^{-2}		
9.	NH_4^+		
10.		H_2O	

11. Which is a stronger base, HSO_4^- or $H_2PO_4^-$? _____

12. Which is a weaker base, Cl^- or NO_2^- ? _____

pH and pOH

The pH of a solution indicates how acidic or basic that solution is. A pH of less than 7 is acidic; a pH of 7 is neutral; and a pH greater than 7 is basic.

Since $(H^+)(OH^-) = 10^{-14}$ at 25°C, if (H^+) is known, the (OH^-) can be calculated and vice versa.

$pH = -\log (H^+)$ So if $(H^+) = 10^{-6}$ M, pH = 6.

$pOH = -\log (OH^-)$ So if $(OH^-) = 10^{-8}$ M, pOH = 8.

Together, $pH + pOH = 14$.

Complete the chart.

	(H^+)	pH	(OH^-)	pOH	Acidic or Basic
1.	10^{-5} M	5	10^{-9} M	9	acidic
2.		7			
3.			10^{-4} M		
4.	10^{-2} M				
5.				11	
6.		12			
7.			10^{-5} M		
8.	10^{-11} M				
9.				13	
10.		6			

pH of Solutions

Calculate the pH of each solution.

1.   0.01 M HCl
2.   0.0010 M NaOH
3.   0.050 M Ca(OH)$_2$
4.   0.030 M HBr
5.   0.150 M KOH
6.   2.0 M HC$_2$H$_3$O$_2$ (Assume 5.0% dissociation.)
7.   3.0 M HF   (Assume 10.0% dissociation.)
8.   0.50 M HNO$_3$
9.   2.50 M NH$_4$OH   (Assume 5.00% dissociation.)
10. 5.0 M HNO$_2$ (Assume 1.0% dissociation.)

Acid-Base Titration

To determine the concentration of an acid (or base), we can react it with a base (or acid) of known concentration until it is completely neutralized. This point of exact neutralization, known as the **endpoint**, is noted by the change in color of the indicator.

Use the following equation:

$$N_A \times V_A = N_B \times V_B$$

where N = normality
V = volume

Solve each problem.

1.	A 25.0 mL sample of HCl was titrated to the endpoint with 15.0 mL of 2.0 N NaOH. What was the normality of the HCl?
2.	A 10.0 mL sample of H_2SO_4 was exactly neutralized by 13.5 mL of 1.0 M KOH. What is the normality of the H_2SO_4?
3.	How much 1.5 M NaOH is necessary to exactly neutralize 20.0 mL of 2.5 M H_3PO_4?
4.	How much 0.5 M HNO_3 is necessary to titrate 25.0 mL of 0.05 M $Ca(OH)_2$ solution to the endpoint?
5.	What is the molarity of a NaOH solution if 15.0 mL is exactly neutralized by 7.5 mL of a 0.02 M $HC_2H_3O_2$ solution?

Hydrolysis of Salts

Salt solutions may be acidic, basic, or neutral, depending on the original acid and base that formed the salt.

  strong acid + strong base → neutral salt

  strong acid + weak base → acidic salt

  weak acid + strong base → basic salt

A weak acid and a weak base will produce any type of solution depending on the relative strengths of the acid and base involved.

Complete the chart for each salt shown.

Salt	Parent Acid	Parent Base	Type of Solution
1. KCl			
2. NH_4NO_3			
3. Na_3PO_4			
4. $CaSO_4$			
5. $AlBr_3$			
6. CuI_2			
7. MgF_4			
8. $NaNO_3$			
9. $LiC_2H_3O_2$			
10. $ZnCl_2$			
11. $SrSO_4$			
12. $Ba_3(PO_4)_2$			

Acids and Bases Crossword

Across

1. Scale of acidity
5. An acid that consists of only two elements
7. Substance that forms hydronium ions in water (Arrhenius _____)
8. This happens when an acid dissolves in water.
10. According to Bronsted-Lowry, an acid is a _____ donor.
12. According to Bronsted-Lowry, a base is a proton _____.
13. Can act as either an acid or a base
14. These pairs differ only by a proton.
16. An acid with a small K_a value would be a _____ acid.
17. Reaction of an ion with H_2O to produce $H^+(aq) + OH^-(aq)$

Down

2. H_3O^+
3. Formed from the reaction of an acid and a base
4. Procedure to determine the concentration of an acid or base
6. A solution that will resist changes in pH
8. Changes color at the endpoint of a titration
9. The reaction of an acid with a base
11. Substance that produces hydroxide ions in aqueous solution (Arrhenius _____)
15. When equivalent amounts of H^+ and OH^- have reacted in a titration

Assigning Oxidation Numbers

Assign oxidation numbers to all of the elements in each compound or ion shown.

1. HCl	11. H_2SO_3
2. KNO_3	12. H_2SO_4
3. OH^-	13. BaO_2
4. Mg_3N_2	14. $KMnO_4$
5. $KClO_3$	15. LiH
6. $Al(NO_3)_3$	16. MnO_2
7. S_8	17. OF_2
8. H_2O_2	18. SO_3
9. PbO_2	19. NH_3
10. $NaHSO_4$	20. Na

Redox Reactions

For each equation, identify the substance oxidized, the substance reduced, the oxidizing agent, and the reducing agent. Then, write the oxidation and reduction half-reactions.

Example:

oxidized   reduced

$$Mg \; + \; Br_2 \; \rightarrow \; MgBr_2$$

reducing   oxidizing
agent   agent

oxidation half-reaction: $Mg^\circ \rightarrow Mg^{+2} + 2e^-$

reduction half-reaction: $2e^- + Br_2^\circ \rightarrow 2Br^-$

1. $2H_2 \; + \; O_2 \; \rightarrow \; 2H_2O$

2. $Fe \; + \; Zn^{2+} \; \rightarrow \; Fe^{2+} \; + \; Zn$

3. $2Al \; + \; 3Fe^{2+} \; \rightarrow \; 2Al^{3+} \; + \; 3Fe$

4. $Cu \; + \; 2AgNO_3 \; \rightarrow \; Cu(NO_3)_2 \; + \; 2Ag$

Balancing Redox Equations

Balance each equation using the half-reaction method.

1. Sn° + Ag^+ → Sn^{2+} + Ag°

2. Cr° + Pb^{2+} → Cr^{3+} + Pb°

3. $KClO_3$ → KCl + O_2

4. NH_3 + O_2 → NO + H_2O

5. PbS + H_2O_2 → $PbSO_4$ + H_2O

6. H_2S + HNO_3 → S + NO + H_2O

7. MnO_2 + $H_2C_2O_4$ + H_2SO_4 → $MnSO_4$ + CO_2 + H_2O

8. H_2S + H_2SO_3 → S + H_2O

9. KIO_3 + H_2SO_3 → KI + H_2SO_4

10. $K_2Cr_2O_7$ + HCl → KCl + $CrCl_3$ + Cl_2 + H_2O

The Electrochemical Cell

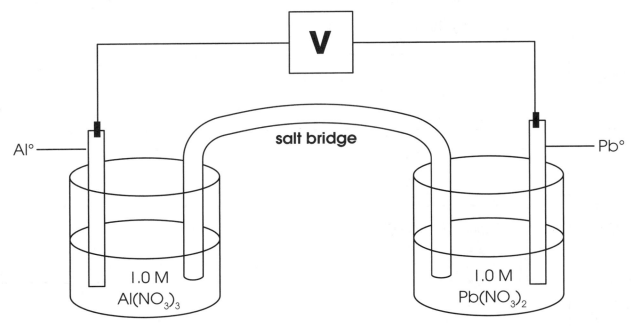

Answer each question, referring to the diagram and a table of standard electrode potentials.

1. Which is the more easily oxidized metal: aluminum or lead? _____

2. What is the balanced equation showing the spontaneous reaction that occurs?

   _____

3. What is the maximum voltage that the above cell can produce? _____

4. What is the direction of electron flow in the wire? _____

5. What is the direction of positive ion flow in the salt bridge? _____

6. Which electrode is decreasing in size? _____

7. Which electrode is increasing in size? _____

8. What is happening to the concentration of aluminum ions? _____

9. What is happening to the concentration of lead ions? _____

10. What is the voltage in this cell when the reaction reaches equilibrium? _____

11. Which is the anode? _____

12. Which is the cathode? _____

13. Which is the positive electrode? _____

14. Which is the negative electrode? _____

Name_____

Electrochemistry Crossword

ACROSS

4. Unit of electrical potential
6. Electrode where oxidation tokes place
7. Both atoms and _____ must be balanced in a redox equation.
9. The anode in an electrochemical cell has this charge.
10. Gain of electrons
12. Voltage of an electrochemical cell when it reaches equilibrium
13. A substance that is oxidized is the _____ agent.
14. Allows the flow of ions in an electrochemical cell

DOWN

1. The anode in an electrolytic cell has this charge.
2. Another word for on electrochemical cell
3. Electrode where reduction takes place
5. Process of layering a metal onto a surface in an electrolytic cell
8. Loss of electrons
11. A substance that is reduced is the _____ agent.

Naming Hydrocarbons

Name each compound according to the IUPAC naming system.

1.

```
    H   H   H
    |   |   |
H — C — C — C — H
    |   |   |
    H   H   H
```

2.

```
    H   H   H   H
    |   |   |   |
H — C = C — C — C — H
            |   |
            H   H
```

3.

```
                H
                |
H — C ≡ C — C — H
                |
                H
```

4.

```
    H   H   H   CH₃ H
    |   |   |   |   |
H — C — C — C — C — C — H
    |   |   |   |   |
    H   H   H   H   H
```

5.

```
    H   H   H
    |   |   |
H — C — C — C — H
    |   |   |
    H   H   H — C — H
                |
                H
```

6.

```
    H   CH₃ H
    |   |   |
H — C — C — C — H
    |   |   |
    H   H   H
```

7.

```
    H   H   H   H   H
    |   |   |   |   |
H — C — C = C — C — C — H
                |   |
                H   H
```

8.

```
                H
                |
            H — C — H
                |
            H — C — H
    H   H   |       H   H
    |   |   |       |   |
H — C — C — C — C — C — H
    |   |   |       |   |
    H   H   |       H   H
            H — C — H
                |
            H   C   H
                |
                H
```

Structure of Hydrocarbons

Draw the structure of each compound.

1. ethane	5. ethyne
2. propene	6. 3, 3-dimethylpentane
3. 2-butene	7. 2, 3-dimethylpentane
4. methane	8. n-butyne

Functional Groups

Classify each of the organic compounds as an *alcohol, carboxylic acid, aldehyde, ketone, ether,* or *ester*. Then, draw its structural formula.

1. CH_3COOH	6. $CH_3CH(OH)CH_3$
2. CH_3COCH_3	7. CH_3CH_2COOH
3. CH_3CH_2OH	8. $CH_3CH_2COOCH_3$
4. $CH_3CH_2OCH_3$	9. $CH_3CH_2COCH_3$
5. CH_3CH_2CHO	10. CH_3OCH_3

Naming Other Organic Compounds

Name each compound.

1.

```
      H   H
      |   |
  H — C — C — OH
      |   |
      H   H
```

2.

```
      H   O   H
      |   ‖   |
  H — C — C — C — H
      |       |
      H       H
```

3.

```
      H   H   H   O
      |   |   |   ‖
  H — C — C — C — C — H
      |   |   |
      H   H   H
```

4.

```
      H   O
      |   ‖
  H — C — C — OH
      |
      H
```

5.

```
      H       H
      |       |
  H — C — O — C — H
      |       |
      H       H
```

6.

```
      H   O       H
      |   ‖       |
  H — C — C — O — C — H
      |           |
      H           H
```

7.

```
      H   OH  H   H
      |   |   |   |
  H — C — C — C — C — H
      |   |   |   |
      H   H   H   H
```

8.

```
      H   H   O
      |   |   ‖
  H — C — C — C — OH
      |   |
      H   H
```

9.

```
      H   O
      |   ‖
  H — C — C — H
      |
      H
```

10.

```
      H   H   O   H
      |   |   ‖   |
  H — C — C — C — C — H
      |   |       |
      H   H       H
```

Structures of Other Organic Compounds

Draw the structure of each compound.

1. butanoic acid	6. methyl methanoate (methyl formate)
2. methanal	7. 3-pentanol
3. methanol	8. methanoic acid (formic acid)
4. butanone	9. propanal
5. diethyl ether	10. 2-pentanone

Name_____

Organic Chemistry Crossword

Across

2. Hydrocarbons containing only single bonds

4. Contains two double bonds

5. Alcohol in which the hydroxyl group is attached to an end carbon

7. An alkane minus one hydrogen; It attaches to another carbon chain.

9. Compounds with the same molecular formula, but different structural formulas

11. A dihydroxy alcohol

12. Alcohol in which the hydroxyl is attached to a carbon attached to two other carbons

14. Open chain hydrocarbon containing one double bond

17. Organic compounds containing the benzene ring structure

18. Describes a hydrocarbon with a side chain of carbon atoms

Down

1. An alcohol with only one hydroxyl group in its structure

3. Alcohol in which the hydroxyl is attached to a carbon attached to three other carbons

4. General formula R-CHO

6. High molecular mass compound consisting of repeating units called monomers

8. General formula R-CO-R'

10. Contains one or more -OH groups

13. Saturated open chain hydrocarbon

14. Open chain hydrocarbon containing only one triple bond

15. Produced by the reaction of an alcohol and an acid

16. General formula R-O-R'

Reference

Selected Polyatomic Ions

Hg_2^{2+}	dimercury(I)	CrO_4^{2-}	chromate
NH_4^+	ammonium	$Cr_2O_7^{2-}$	dichromate
$C_2H_3O_2^-$ $CH_3COO\text{-}$ } acetate		MnO_4^-	permanganate
		MnO_4^{2-}	manganate
CN^-	cyanide	NO_2^-	nitrite
CO_3^{2-}	carbonate	NO_3^-	nitrate
HCO_3^-	hydrogen carbonate	OH^-	hydroxide
$C_2O_4^{2-}$	oxalate	PO_4^{3-}	phosphate
ClO^-	hypochlorite	SCN	thiocyanate
ClO_2^-	chlorite	SO_3^{2-}	sulfite
ClO_3^-	chlorate	SO_4^{2-}	sulfate
ClO_4^-	perchlorate	HSO_4^-	hydrogen sulfate
		$S_2O_3^{2-}$	thiosulfate

Solutions

Boiling Point Elevation	$\Delta T_b = K_b m$
Dilution	$V_1 C_1 = V_2 C_2$
Freezing Point Depression	$\Delta T_f = \Delta T_f = K_f m$
Henry's Law	$C_g = k_g P_g$ or $S = k_H P$
Molality	$m = \dfrac{\text{mol solute}}{\text{kg solvent}}$
Molarity	$M = \dfrac{\text{mol solute}}{\text{L solution}}$
Normality	$N = \dfrac{\text{equivilants solute}}{\text{L solution}}$
Raoult's Law	$P_A = P°_A X_A$ $\Delta P_A = P°_A X_B$

Fundamental Concepts

Density	$D = \dfrac{m}{V}$
Force	$F = ma$
Pressure	$P = \dfrac{F}{A}$
Temperature Conversion	$°C = \dfrac{5}{9}(°F - 32)$ $°F = \left(\dfrac{5}{9}\right)°C + 32$
Work	$w = Fd$

Gas Laws

Boyle's Law	$P_1 V_1 = P_2 V_2$
Charles' Law	$V_1 T_2 = V_2 T_1$
Combined Gas Law	$V_1 P_1 T_2 = V_2 P_2 T_1$
Dalton's Law	$P_T = P_1 + P_2 + P_3 \ldots$
Graham's Law	$\dfrac{r_1}{r_2} = \dfrac{u_1}{u_2} = \sqrt{\dfrac{3RT/M_1}{3RT/M_2}} = \sqrt{\dfrac{M_2}{M_1}}$
Ideal Gas Law	$PV = nRT$
Van der Waals Equation	$\left(P + \dfrac{an^2}{V^2}\right)(V - nb) = nRT$

Energy Changes in Chemical Reactions

Enthalpy	$\Delta H = H_{products} - H_{reactants}$
	$\Delta E = q + w = q - P\Delta V$
	$qp = \Delta E + P\Delta V = \Delta H$
Entropy	$\Delta S = S_{products} - S_{reactants}$
Free Energy	$\Delta G° = \Delta H° - T(\Delta S°)$

Chemical Equilibrium

Equilibrium Constant Expression

for $aA + bB \rightleftharpoons cC + dD$

$$K_c = \dfrac{(C)^c (D)^d}{(A)^a (B)^b} = \dfrac{k_{forward\ reaction}}{k_{reverse\ reaction}}$$

Electrochemistry

Electromotive Force	$E° = \left(\dfrac{2.303RT}{n\mathcal{F}}\right)\log K$
Nernst Equation	$E = E° - \left(\dfrac{0.0591}{n}\right)\log Q$

Acidity and Basicity

Ion Product Constant of Water	$k_w = (H^+)(OH^-) = 1.0 \times 10^{-14}$
pH	$pH = -\log(H^+)$ $pOH = -\log(OH^-)$
	$pH + pOH = -\log K_w = 14.00$
	$K_a K_b = (H^+)(OH^-) = k_w$

Radiant Energy

Wave Velocity	$c = \nu\lambda$
Energy of a Photon	$E = h\nu$

Ionic Equilibrium

Henderson-Hasselbach Equation	$pH = pK_a + \log\dfrac{(base)}{(acid)}$
Solubility Product Constant	$K_{sp} = (A^+)^m (X^-)^n$

Behavior of Nuclear Particles

Mass-Energy Relationship	$E = mc^2$
Wavelength of a Particle	$\lambda = \dfrac{h}{m\nu}$

Answer Key

Laboratory Dos And Don'ts

Identify what is wrong in each laboratory activity.

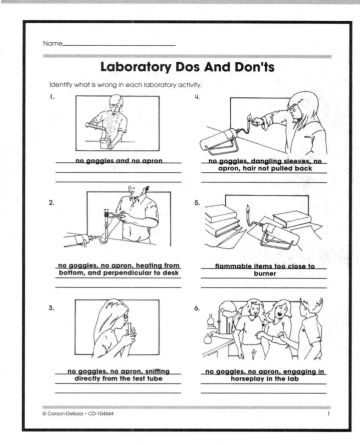

1. no goggles and no apron

2. no goggles, no apron, heating from bottom, and perpendicular to desk

3. no goggles, no apron, sniffing directly from the test tube

4. no goggles, dangling sleeves, no apron, hair not pulled back

5. flammable items too close to burner

6. no goggles, no apron, engaging in horseplay in the lab

© Carson-Dellosa • CD-104644   1

Laboratory Equipment

Label the lab equipment.

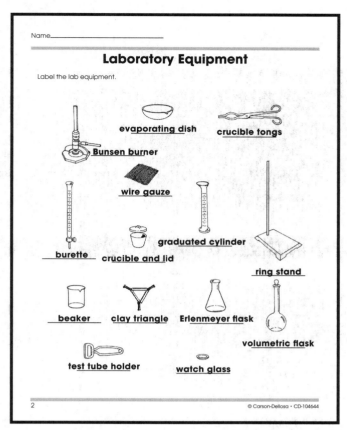

evaporating dish   crucible tongs

Bunsen burner

wire gauze

graduated cylinder

burette   crucible and lid

ring stand

beaker   clay triangle   Erlenmeyer flask

volumetric flask

test tube holder   watch glass

2   © Carson-Dellosa • CD-104644

Triple And Four Beam Balances

Identify the mass on each balance.

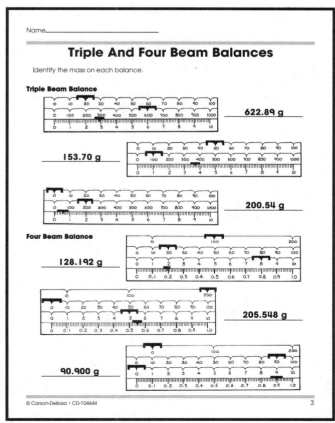

Triple Beam Balance

622.89 g

153.70 g

200.54 g

Four Beam Balance

128.192 g

205.548 g

90.900 g

© Carson-Dellosa • CD-104644   3

Measuring Liquid Volume

Identify the volume indicated on each graduated cylinder. The unit of volume is mL.

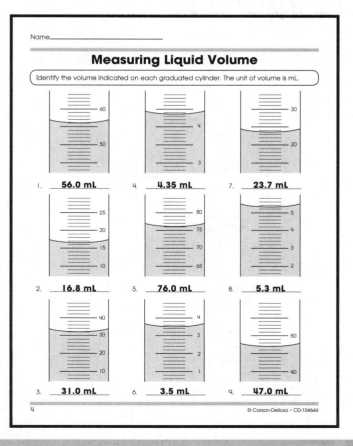

1. 56.0 mL   4. 4.35 mL   7. 23.7 mL

2. 16.8 mL   5. 76.0 mL   8. 5.3 mL

3. 31.0 mL   6. 3.5 mL   9. 47.0 mL

4   © Carson-Dellosa • CD-104644

Answer Key

Name_____

Reading Thermometers

Identify the temperature indicated on each thermometer.

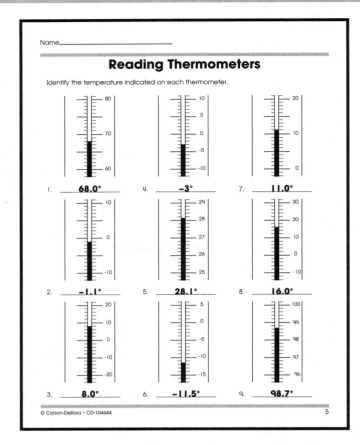

1. **68.0°**   4. **–3°**   7. **11.0°**

2. **–1.1°**   5. **28.1°**   8. **16.0°**

3. **8.0°**   6. **–11.5°**   9. **98.7°**

© Carson-Dellosa • CD-104644   5

Name_____

Metrics And Measurements

In the chemistry classroom and lab, the metric system of measurement is used. It is important to be able to convert from one unit to another.

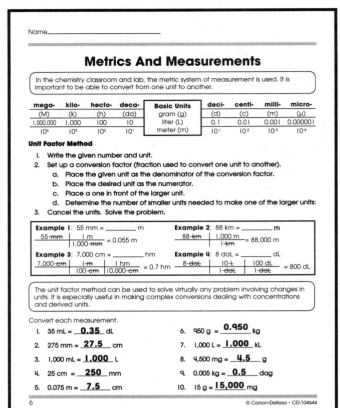

mega-	kilo-	hecto-	deca-	Basic Units	deci-	centi-	milli-	micro-
(M)	(k)	(h)	(da)	gram (g)	(d)	(c)	(m)	(µ)
1,000,000	1,000	100	10	liter (L)	0.1	0.01	0.001	0.000001
10^6	10^3	10^2	10^1	meter (m)	10^{-1}	10^{-2}	10^{-3}	10^{-6}

Unit Factor Method

1. Write the given number and unit.
2. Set up a conversion factor (fraction used to convert one unit to another).
   a. Place the given unit as the denominator of the conversion factor.
   b. Place the desired unit as the numerator.
   c. Place a one in front of the larger unit.
   d. Determine the number of smaller units needed to make one of the larger units:
3. Cancel the units. Solve the problem.

Example 1: 55 mm = _____ m
$$\frac{55\ mm}{} \times \frac{1\ m}{1,000\ mm} = 0.055\ m$$

Example 2: 88 km = _____ m
$$\frac{88\ km}{} \times \frac{1,000\ m}{1\ km} = 88,000\ m$$

Example 3: 7,000 cm = _____ hm
$$\frac{7,000\ cm}{} \times \frac{1\ m}{100\ cm} \times \frac{1\ hm}{10,000\ cm} = 0.7\ hm$$

Example 4: 8 daL = _____ dL
$$\frac{8\ daL}{} \times \frac{10\ L}{1\ daL} \times \frac{100\ dL}{1\ daL} = 800\ dL$$

The unit factor method can be used to solve virtually any problem involving changes in units. It is especially useful in making complex conversions dealing with concentrations and derived units.

Convert each measurement.

1. 35 mL = **0.35** dL
2. 275 mm = **27.5** cm
3. 1,000 mL = **1.000** L
4. 25 cm = **250** mm
5. 0.075 m = **7.5** cm

6. 950 g = **0.950** kg
7. 1,000 L = **1.000** kL
8. 4,500 mg = **4.5** g
9. 0.005 kg = **0.5** dag
10. 15 g = **15,000** mg

6   © Carson-Dellosa • CD-104644

Name_____

Dimensional Analysis (Unit Factor Method)

Using this method, it is possible to solve many problems by using the relationship of one unit to another. For example, 12 inches = one foot. Since these two numbers represent the same value, the fractions 12 in./1 ft. and 1 ft./12 in. are both equal to one. When you multiply another number by the number one, you do not change its value. However, you may change its unit.

Example 1: Convert 2 miles to inches.
$$2\ miles \times \frac{5,280\ ft.}{1\ mile} \times \frac{12\ inches}{1\ ft.} = 126,720\ in.$$

Example 2: How many seconds are in 4 days?
$$4\ days \times \frac{24\ hrs.}{1\ day} \times \frac{60\ min.}{1\ hr.} \times \frac{60\ sec.}{1\ min.} = 345,600\ sec.$$

Solve each problem. Round irrational numbers to the thousandths place.

1. 3 hr. = **10,800** sec.
2. 0.035 mg = **0.0035** cg
3. 5.5 kg = **12.1** lb.
4. 2.5 yd. = **90** in.
5. 1.3 yr. = **28,470** hr.
6. 3 moles = **1.806×10^{24}** molecules (1 mole = 6.02×10^{23} molecules)
7. 2.5×10^{24} molecules = **4.152** moles
8. 5 moles = **112** liters (1 mole = 22.4 liters)
9. 100. liters = **4.46** moles
10. 50. liters = **1.344×10^{24}** molecules
11. 5.0×10^{24} molecules = **186.047** liters
12. 7.5×10^3 mL = **7.5** liters

© Carson-Dellosa • CD-104644   7

Name_____

Scientific Notation

Scientists very often deal with very small and very large numbers, which can lead to a lot of confusion when counting zeros. We can express these numbers as powers of 10.

Scientific notation takes the form of $M \times 10^n$ where $1 \le M < 10$ and n represents the number of decimal places to be moved. Positive n indicates the standard form is a large number. Negative n indicates a number between zero and one.

Example 1: Convert 1,500,000 to scientific notation.
Move the decimal point so that there is only one digit to its left, for a total of 6 places.
$$1,500,000 = 1.5 \times 10^6$$

Example 2: Convert 0.000025 to scientific notation.
For this, move the decimal point 5 places to the right.
$$0.000025 = 2.5 \times 10^{-5}$$

(Note that when a number starts out less than one, the exponent is always negative.)

Convert each number to scientific notation.

1. 0.005 = **5×10^{-3}**
2. 5,050 = **5.05×10^3**
3. 0.0008 = **8×10^{-4}**
4. 1,000 = **1×10^3**
5. 1,000,000 = **1×10^6**

6. 0.25 = **2.5×10^{-1}**
7. 0.025 = **2.5×10^{-2}**
8. 0.0025 = **2.5×10^{-3}**
9. 500 = **5×10^2**
10. 5,000 = **5×10^3**

Convert each number to standard notation.

11. 1.5×10^3 = **1,500**
12. 1.5×10^{-3} = **0.0015**
13. 3.75×10^{-2} = **0.0375**
14. 3.75×10^2 = **375**
15. 2.2×10^5 = **220,000**

16. 3.35×10^{-1} = **0.335**
17. 1.2×10^{-4} = **0.00012**
18. 1×10^4 = **10,000**
19. 1×10^{-1} = **0.1**
20. 4×10^0 = **4**

8   © Carson-Dellosa • CD-104644

© Carson-Dellosa • CD-104644

Answer Key

Name_____

Significant Figures

A measurement can only be as accurate and precise as the instrument that produced it. A scientist must be able to express the accuracy of a number, not just its numerical value. We can determine the accuracy of a number by the number of significant figures it contains.

1. All digits 1–9 inclusive are significant.

    Example: 129 has 3 significant figures.

2. Zeros between significant digits are always significant.

    Example: 5,007 has 4 significant figures.

3. Trailing zeros in a number are significant only if the number contains a decimal point. Sometimes, a decimal may be added without any number in the tenths place.

    Example: 100.0 has 4 significant figures.

    100. has 3 significant figures.

    100 has 1 significant figure.

4. Zeros in the beginning of a number whose only function is to place the decimal point are not significant.

    Example: 0.0025 has 2 significant figures.

5. Zeros following a decimal significant figure are significant.

    Example: 0.000470 has 3 significant figures.

    0.47000 has 5 significant figures.

Determine the number of significant figures in each number.

1. 0.02 __**1**__
2. 0.020 __**2**__
3. 501 __**3**__
4. 501.0 __**4**__
5. 5,000 __**1**__
6. 5,000. __**4**__
7. 6,051.00 __**6**__
8. 0.0005 __**1**__
9. 0.1020 __**4**__
10. 10,001 __**5**__

Determine the location of the last significant place value by placing a bar over the digit. (Example: 1.70$\overline{0}$)

11. 8,0$\overline{4}$0
12. 0.9010$\overline{0}$
13. 3.01 × 10$^{\overline{2}1}$
14. 0.030$\overline{0}$
15. 90,$\overline{1}$00
16. 0.000041$\overline{0}$
17. 699.$\overline{5}$
18. 4.$\overline{7}$ × 10^{-8}
19. 2.00$\overline{0}$ × 10^2
20. 10,800,0$\overline{0}$0.0

Name_____

Calculations Using Significant Figures

When multiplying and dividing, limit and round to the least number of significant figures in any of the factors.

Example 1: 23.0 cm × 432 cm × 19 cm = 188,784 cm³

    The answer is expressed as 190,000 cm³ since 19 cm has only two significant figures.

When adding and subtracting, limit and round your answer to the least number of decimal places in any of the numbers that make up your answer.

Example 2: 123.25 mL + 46.0 mL + 86.257 mL = 255.507 mL

    The answer is expressed as 255.5 mL since 46.0 mL has only one decimal place.

Perform each operation, expressing the answer in the correct number of significant figures.

1. 1.35 m × 2.467 m = __**3.33 m²**__

2. 1,035 m² ÷ 42 m = __**25 m**__

3. 12.01 mL + 35.2 mL + 6 mL = __**53 mL**__

4. 55.46 g – 28.9 g = __**26.6 g**__

5. 0.021 cm × 3.2 cm × 100.1 cm = __**6.7 cm³**__

6. 0.15 cm + 1.15 cm + 2.051 cm = __**3.35 cm**__

7. 150 L³ ÷ 4 L = __**40 L²**__

8. 505 kg – 450.25 kg = __**55 kg**__

9. 1.252 mm × 0.115 mm × 0.012 mm = __**0.0017 mm³**__

10. 1.278 × 10³ m² ÷ 1.4267 × 10² m = __**8.958 m**__

Name_____

Percentage Error

Percentage error is a way for scientists to express how far off a laboratory value is from the commonly accepted value.

The formula is:

$$\% \text{ error} = \left| \frac{\text{Accepted Value} - \text{Experimental Value}}{\text{Accepted Value}} \right| \times 100$$

absolute value

Determine the percentage error in each problem.

1. Experimental value = 1.24 g
   Accepted value = 1.30 g   **4.62%**

2. Experimental value = 1.24 × 10⁻² g
   Accepted value = 9.98 × 10⁻³ g   **24.2%**

3. Experimental value = 252 mL
   Accepted value = 225 mL   **12.0 %**

4. Experimental value = 22.2 L
   Accepted value = 22.4 L   **0.893%**

5. Experimental value = 125.2 mg
   Accepted value = 124.8 mg   **0.3%**

Name_____

Temperature and Its Measurement

Temperature (which measures average kinetic energy of the molecules) can be measured using three common scales: **Celsius, Kelvin,** and **Fahrenheit.** Use the following formulas to convert from one scale to another. Celsius is the scale most desirable for laboratory work. Kelvin represents the absolute scale. Fahrenheit is the old English scale, which is rarely used in laboratories.

$$°C = K - 273 \qquad K = °C + 273$$
$$°F = \frac{9}{5}°C + 32 \qquad °C = \frac{5}{9}(°F - 32)$$

Complete the chart. All measurements are good to 1°C or better.

	°C	K	°F
1.	0°C	**273 K**	**32°F**
2.	**100°C**	**373 K**	212°F
3.	**177°C**	450 K	**351°F**
4.	**37.0°C**	**310 K**	98.6°F
5.	–273°C	**0 K**	**–459°F**
6.	**21°C**	294 K	**70°F**
7.	**25°C**	**298 K**	77°F
8.	**–48°C**	225 K	**–54°F**
9.	–40°C	**233 K**	**–40°F**

Answer Key

Name_____

Freezing and Boiling Point Graph

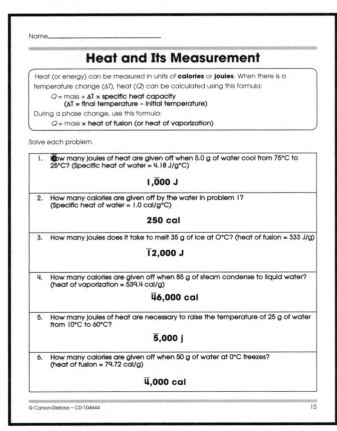

Use the graph to answer each question.

1. Which is the freezing point of the substance? **5°C**
2. Which is the boiling point of the substance? **15°C**
3. Which is the melting point of the substance? **5°C**
4. Which letter represents the range where the solid is being warmed? **A**
5. Which letter represents the range where the liquid is being warmed? **C**
6. Which letter represents the range where the vapor is being warmed? **E**
7. Which letter represents the melting of the solid? **B**
8. Which letter represents the vaporization of the liquid? **D**
9. Which letter(s) shows a change in potential energy? **B, D**
10. Which letter(s) shows a change in kinetic energy? **A, C, E**
11. Which letter represents condensation? **D**
12. Which letter represents crystallization? **B**

© Carson-Dellosa • CD-104644

13

Name_____

Phase Diagram

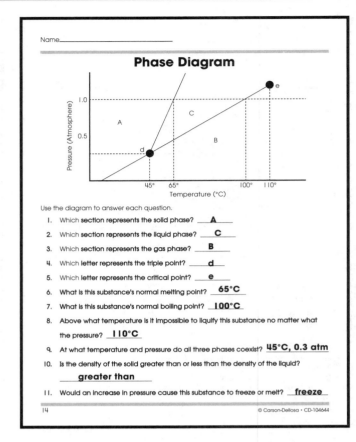

Use the diagram to answer each question.

1. Which **section** represents the solid phase? **A**
2. Which **section** represents the liquid phase? **C**
3. Which **section** represents the gas phase? **B**
4. Which **letter** represents the triple point? **d**
5. Which **letter** represents the critical point? **e**
6. What is this substance's normal melting point? **65°C**
7. What is this substance's normal boiling point? **100°C**
8. Above what temperature is it impossible to liquify this substance no matter what the pressure? **110°C**
9. At what temperature and pressure do all three phases coexist? **45°C, 0.3 atm**
10. Is the density of the solid greater than or less than the density of the liquid? **greater than**
11. Would an increase in pressure cause this substance to freeze or melt? **freeze**

14
© Carson-Dellosa • CD-104644

Name_____

Heat and Its Measurement

Heat (or energy) can be measured in units of **calories** or **joules**. When there is a temperature change (ΔT), heat (Q) can be calculated using this formula:

Q = mass × ΔT × specific heat capacity
(ΔT = final temperature – initial temperature)

During a phase change, use this formula:

Q = mass × **heat of fusion (or heat of vaporization)**

Solve each problem.

1. How many joules of heat are given off when 5.0 g of water cool from 75°C to 25°C? (Specific heat of water = 4.18 J/g°C)

   1,000 J

2. How many calories are given off by the water in problem 1? (Specific heat of water = 1.0 cal/g°C)

   250 cal

3. How many joules does it take to melt 35 g of ice at 0°C? (heat of fusion = 333 J/g)

   12,000 J

4. How many calories are given off when 85 g of steam condense to liquid water? (heat of vaporization = 539.4 cal/g)

   46,000 cal

5. How many joules of heat are necessary to raise the temperature of 25 g of water from 10°C to 60°C?

   5,000 j

6. How many calories are given off when 50 g of water at 0°C freezes? (heat of fusion = 79.72 cal/g)

   4,000 cal

© Carson-Dellosa • CD-104644

15

Name_____

Vapor Pressure and Boiling

A liquid will boil when its vapor pressure equals the atmospheric pressure.

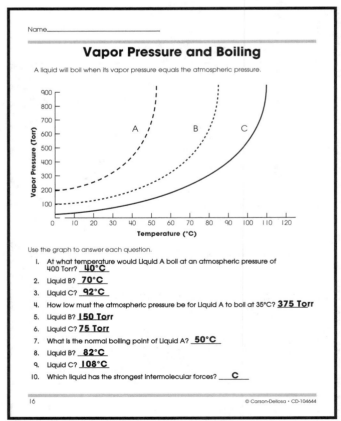

Use the graph to answer each question.

1. At what temperature would Liquid A boil at an atmospheric pressure of 400 Torr? **40°C**
2. Liquid B? **70°C**
3. Liquid C? **92°C**
4. How low must the atmospheric pressure be for Liquid A to boil at 35°C? **375 Torr**
5. Liquid B? **150 Torr**
6. Liquid C? **75 Torr**
7. What is the normal boiling point of Liquid A? **50°C**
8. Liquid B? **82°C**
9. Liquid C? **108°C**
10. Which liquid has the strongest intermolecular forces? **C**

16
© Carson-Dellosa • CD-104644

© Carson-Dellosa • CD-104644

Answer Key

Name_____

Matter—Substances vs. Mixtures

All matter can be classified as either a **substance** (element or compound) or a **mixture** (heterogeneous or homogeneous).

Matter
- Substance — can write chemical formula, homogeneous
  - Element — one type of atom
  - Compound — two or more different atoms, chemically bonded
- Mixture — variable ratio
  - Homogeneous — solutions
  - Heterogeneous — colloids and suspensions

Classify each of the following as a substance or a mixture. If it is a substance, write *element* or *compound* in the substance column. If it is a mixture, write *heterogeneous* or *homogeneous* in the mixture column.

Type of Matter	Substance	Mixture
1. chlorine	element	
2. water	compound	
3. soil		heterogeneous
4. sugar water		homogeneous
5. oxygen	element	
6. carbon dioxide	compound	
7. rocky road ice cream		heterogeneous
8. alcohol	compound	
9. pure air		homogeneous
10. iron	element	

Name_____

Physical vs. Chemical Properties

A **physical property** is observed with the senses and can be determined without destroying the object. Color, shape, mass, length, and odor are all examples of physical properties.

A **chemical property** indicates how a substance reacts with something else. The original substance is fundamentally changed in observing a chemical property. For example, the ability of iron to rust is a chemical property. The iron has reacted with oxygen, and the original iron metal is changed. It now exists as iron oxide, a different substance.

Classify each property as either chemical or physical by putting a check in the appropriate column.

	Physical Property	Chemical Property
1. blue color	✓	
2. density	✓	
3. flammability		✓
4. solubility	✓	
5. reacts with acid to form H_2		✓
6. supports combustion		✓
7. sour taste	✓	
8. melting point	✓	
9. reacts with water to form a gas		✓
10. reacts with a base to form water		✓
11. hardness	✓	
12. boiling point	✓	
13. can neutralize a base		✓
14. luster	✓	
15. odor	✓	

Name_____

Physical vs. Chemical Changes

In a **physical change**, the original substance still exists; it only changes in form. In a **chemical change**, a new substance is produced. Energy changes always accompany chemical changes.

Classify each as a *physical* or *chemical* change.

1. Sodium hydroxide dissolves in water. **physical**
2. Hydrochloric acid reacts with potassium hydroxide to produce a salt, water, and heat. **chemical**
3. A pellet of sodium is sliced in two. **physical**
4. Water is heated and changed to steam. **physical**
5. Potassium chlorate decomposes to potassium chloride and oxygen gas. **chemical**
6. Iron rusts. **chemical**
7. When placed in H_2O, a sodium pellet catches on fire as hydrogen gas is liberated and sodium hydroxide forms. **chemical**
8. Water evaporates. **physical**
9. Ice melts. **physical**
10. Milk sours. **chemical**
11. Sugar dissolves in water. **physical**
12. Wood rots. **chemical**
13. Pancakes are cooking on a griddle. **chemical**
14. Grass is growing in a lawn. **chemical**
15. A tire is inflated with air. **physical**
16. Food is digested in the stomach. **chemical**
17. Water is absorbed by a paper towel. **physical**

Name_____

Boyle's Law

Boyle's Law states that the volume of a given sample of gas at a constant temperature varies inversely with the pressure. (If one goes up, the other goes down.) Use the formula:

$$P_1 \times V_1 = P_2 \times V_2$$

Solve each problem (assuming constant temperature).

1. A sample of oxygen gas occupies a volume of 250. mL at 740. Torr. What volume will it occupy at 800. Torr pressure?

   $23\overline{1}$ mL

2. A sample of carbon dioxide occupies a volume of 3.50 liters at 125 kPa pressure. What pressure would the gas exert if the volume was decreased to 2.00 liters?

   $2\overline{1}9$ kPa

3. A 2.0 liter container of nitrogen has a pressure of 3.2 atm. What volume would be necessary to decrease the pressure to 1.0 atm?

   6.4 L

4. Ammonia gas occupies a volume of 450. mL at a pressure of 720. mmHg. What volume will it occupy at standard pressure?

   $42\overline{6}$ mL

5. A 175 mL sample of neon has its pressure changed from 75 kPa to 150 kPa. What is its new volume?

   $\overline{8}8$ mL

6. A sample of hydrogen at 1.5 atm has its pressure decreased to 0.50 atm, producing a new volume of 750 mL. What was its original volume?

   250 mL

7. Chlorine gas occupies a volume of 1.2 liters at 720 Torr. What volume will it occupy at 1 atm pressure?

   $1.\overline{1}$ L

8. Fluorine gas exerts a pressure of 900. Torr. When the pressure is changed to 1.50 atm, its volume is 250. mL. What was the original volume?

   $3\overline{1}7$ mL

Answer Key

Name_____

Charles' Law

Charles' Law states that the volume of a given sample of gas at a constant pressure is directly proportional to the temperature in Kelvin. Use the following formulas:

$$\frac{V_1}{T_1} = \frac{V_2}{T_2} \quad \text{or} \quad V_1 \times T_2 = V_2 \times T_1$$

$$K = °C + 273$$

Solve each problem (assuming constant pressure).

1. A sample of nitrogen occupies a volume of 250 mL at 25°C. What volume will it occupy at 95°C?
   $3\overline{1}0$ mL

2. Oxygen gas is at a temperature of 40°C when it occupies a volume of 2.3 liters. To what temperature should it be raised to occupy a volume of 6.5 liters?
   $8\overline{8}0$ K or $6\overline{1}0$°C

3. Hydrogen gas was cooled from $1\overline{5}0$°C to $5\overline{0}$°C. Its new volume is 75 mL. What was its original volume?
   $9\overline{8}$ mL

4. Chlorine gas occupies a volume of 25 mL at $3\overline{0}0$ K. What volume will it occupy at 600 K?
   50 mL

5. A sample of neon gas at 50°C and a volume of 2.5 liters is cooled to 25°C. What is the new volume?
   $2.\overline{3}$ L

6. Fluorine gas at $30\overline{0}$ K occupies a volume of $50\overline{0}$ mL. To what temperature should it be lowered to bring the volume to $30\overline{0}$ mL?
   180 K or –93°C

7. Helium occupies a volume of 3.8 liters at -45°C. What volume will it occupy at 45°C?
   5.3 L

8. A sample of argon gas is cooled and its volume went from $38\overline{0}$ mL to $25\overline{0}$ mL. If its final temperature was -55°C, what was its original temperature?
   $33\overline{1}$ K or 58°C

Name_____

Combined Gas Law

In practical terms, it is often difficult to hold any of the variables constant. When there is a change in pressure, volume, and temperature, the combined gas law is used.

$$\frac{P_1 \times V_1}{T_1} = \frac{P_2 \times V_2}{T_2} \quad \text{or} \quad P_1 V_1 T_2 = P_2 V_2 T_1$$

Complete the chart.

	P_1	V_1	T_1	P_2	V_2	T_2
1.	1.5 atm	3.0 L	$2\overline{0}$°C	2.5 atm	**1.9 L**	$3\overline{0}$°C
2.	720 Torr	256 mL	25°C	**$8\overline{0}0$ Torr**	250 mL	$5\overline{0}$°C
3.	$6\overline{0}0$ mmHg	2.5 L	22°C	760 mmHg	1.8 L	**269 K or –4°C**
4.	**1.2 atm**	750 mL	0.0°C	2.0 atm	500 mL	25°C
5.	95 kPa	4.0 L	**295 K or 22°C**	101 kPa	6.0 L	471 K or 198°C
6.	650. Torr	**275 mL**	100°	900. Torr	225 mL	$15\overline{0}$°C
7.	850 mmHg	1.5 L	15°C	**540 mmHg**	2.5 L	$3\overline{0}$°C
8.	125 kPa	125 mL	**544 K or 271°C**	$10\overline{0}$ kPa	$10\overline{0}$ mL	75°C

Name_____

Dalton's Law of Partial Pressures

Dalton's Law says that the sum of the individual pressures of all the gases that make up a mixture is equal to the total pressure, or: $P_t = P_1 + P_2 + P_3 + ...$ The partial pressure of each gas is equal to the mole fraction of each gas times the total pressure.

$$P_t = P_1 + P_2 + P_3 + ... \quad \text{or} \quad \frac{\text{moles gas}_x}{\text{total moles}} \times P_t = P_x$$

Solve each problem.

1. A 250. mL sample of oxygen is collected over water at 25°C and 760.0 Torr. What is the pressure of the dry gas alone? (Vapor pressure of water at 25°C = 23.8 Torr)
   736 Torr

2. A 32.0 mL sample of hydrogen is collected over water at $2\overline{0}$°C and 750.0 Torr. What is the pressure of the dry gas alone? (Vapor pressure of water at $2\overline{0}$°C = 17.5 Torr)
   732.5 Torr

3. A 54.0 mL sample of oxygen is collected over water at 23°C and 770.0 Torr. What is the pressure of the dry gas alone? (Vapor pressure of water at 23°C = 21.1 Torr)
   748.9 Torr

4. A mixture of 2.00 moles of H_2, 3.00 moles of NH_3, 4.00 moles of CO_2, and 5.00 moles of N_2 exerts a total pressure of 800 Torr. What is the partial pressure of each gas?
   H_2 = 114.2 Torr CO_2 = 228.6 Torr
   NH_3 = 171.4 Torr N_2 = 285.7 Torr

5. The partial pressure of F_2 is 300 Torr in a mixture of gases where the total pressure is 1.00 atm. If there are 1.5 total moles in the mixture, how many moles of F_2 are present?
   263 moles

Name_____

Ideal Gas Law

The **ideal gas law** describes the state of an ideal gas. While an ideal gas is hypothetical, the ideal gas law can be used to approximate the behavior of many gases under normal conditions. Use the formula:

$PV = nRT$ where P = pressure in atmospheres
                   V = volume in liters R = Universal Gas Constant
                                                   0.0821 L•atm/mol•K
                   n = number of moles of gas T = Kelvin temperature

Use the ideal gas law to solve each problem.

1. How many moles of oxygen will occupy a volume of 2.5 liters at 1.2 atm and 25°C?
   0.12 moles

2. What volume will 2.0 moles of nitrogen occupy at 720 Torr and $2\overline{0}$°C?
   $5\overline{1}$ L

3. What pressure will be exerted by 25 g of CO_2 at a temperature of 25°C and a volume of $50\overline{0}$ mL?   **28 atm**

4. At what temperature will 5.00 g of Cl_2 exert a pressure of 900. Torr at a volume of $75\overline{0}$ mL?   **154 K or –119°C**

5. What is the density of NH_3 at $80\overline{0}$ Torr and 25°C?   **0.73 g/L**

6. If the density of a gas is 1.2 g/L at 745. Torr and $2\overline{0}$°C, what is its molecular mass?
   29 g/mol

7. How many moles of nitrogen gas will occupy a volume of 347 mL at 6680 Torr and 27°C?   **0.124 moles**

8. What volume will 454 grams (1 lb.) of hydrogen occupy at 1.05 atm and 25°C?
   10,494.6 L

9. Find the number of grams of CO_2 that exert a pressure of 785 Torr at a volume of 32.5 L and a temperature of 32°C.   **$2\overline{3}$ g**

10. An elemental gas has a mass of 10.3 g. If the volume is 58.4 L and the pressure is 758 Torr at a temperature of 2.5°C, what is the gas?   **Br_2 (bromine)**

Answer Key

Answer Key

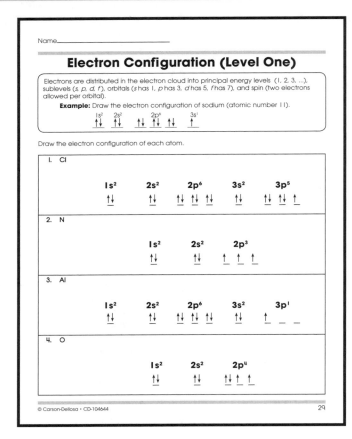

Electron Configuration (Level One)

Electrons are distributed in the electron cloud into principal energy levels (1, 2, 3, ...), sublevels (s, p, d, f), orbitals (s has 1, p has 3, d has 5, f has 7), and spin (two electrons allowed per orbital).

Example: Draw the electron configuration of sodium (atomic number 11).

Draw the electron configuration of each atom.

1. Cl

2. N

3. Al

4. O

© Carson-Dellosa • CD-104644   29

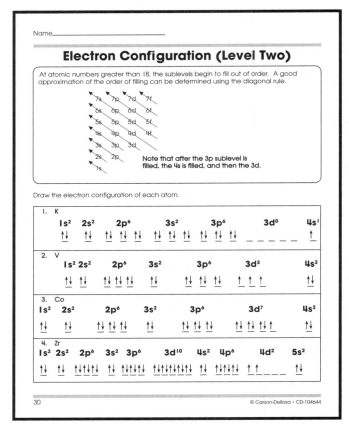

Electron Configuration (Level Two)

At atomic numbers greater than 18, the sublevels begin to fill out of order. A good approximation of the order of filling can be determined using the diagonal rule.

Note that after the 3p sublevel is filled, the 4s is filled, and then the 3d.

Draw the electron configuration of each atom.

1. K
2. V
3. Co
4. Zr

30   © Carson-Dellosa • CD-104644

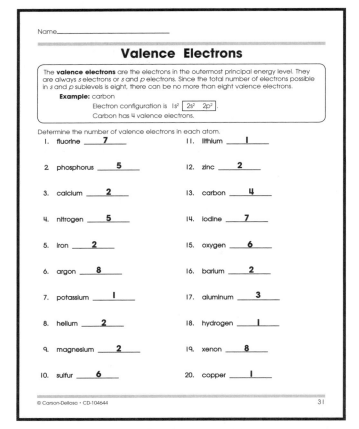

Valence Electrons

The **valence electrons** are the electrons in the outermost principal energy level. They are always s electrons or s and p electrons. Since the total number of electrons possible in s and p sublevels is eight, there can be no more than eight valence electrons.

Example: carbon

Electron configuration is $1s^2$ $2s^2$ $2p^2$.
Carbon has 4 valence electrons.

Determine the number of valence electrons in each atom.

1. fluorine **7**
2. phosphorus **5**
3. calcium **2**
4. nitrogen **5**
5. iron **2**
6. argon **8**
7. potassium **1**
8. helium **2**
9. magnesium **2**
10. sulfur **6**

11. lithium **1**
12. zinc **2**
13. carbon **4**
14. iodine **7**
15. oxygen **6**
16. barium **2**
17. aluminum **3**
18. hydrogen **1**
19. xenon **8**
20. copper **1**

© Carson-Dellosa • CD-104644   31

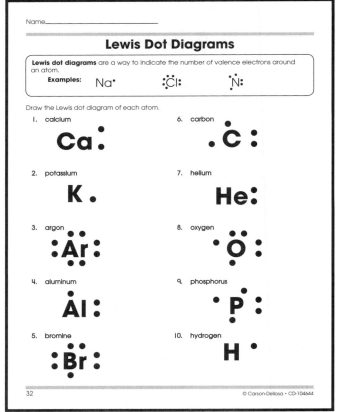

Lewis Dot Diagrams

Lewis dot diagrams are a way to indicate the number of valence electrons around an atom.

Examples: Na•   :C̈l: :N̈:

Draw the Lewis dot diagram of each atom.

1. calcium
2. potassium
3. argon
4. aluminum
5. bromine

6. carbon
7. helium
8. oxygen
9. phosphorus
10. hydrogen

32   © Carson-Dellosa • CD-104644

Answer Key

Atomic Structure Crossword

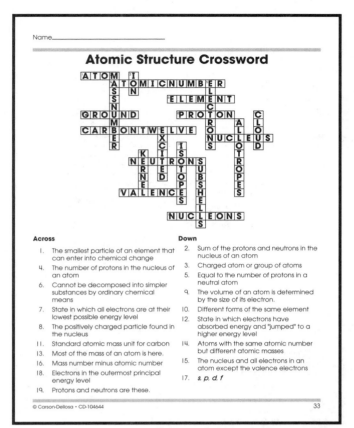

Across

1. The smallest particle of an element that can enter into chemical change
4. The number of protons in the nucleus of an atom
6. Cannot be decomposed into simpler substances by ordinary chemical means
7. State in which all electrons are at their lowest possible energy level
8. The positively charged particle found in the nucleus
11. Standard atomic mass unit for carbon
13. Most of the mass of an atom is here.
16. Mass number minus atomic number
18. Electrons in the outermost principal energy level
19. Protons and neutrons are these.

Down

2. Sum of the protons and neutrons in the nucleus of an atom
3. Charged atom or group of atoms
5. Equal to the number of protons in a neutral atom
9. The volume of an atom is determined by the size of its electron.
10. Different forms of the same element
12. State in which electrons have absorbed energy and "jumped" to a higher energy level
14. Atoms with the same atomic number but different atomic masses
15. The nucleus and all electrons in an atom except the valence electrons
17. *s, p, d, f*

Nuclear Decay

Predict the product of each nuclear reaction.

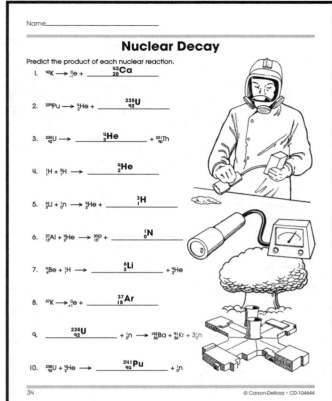

1. $^{42}_{19}K \rightarrow {}^{0}_{-1}e + $ ___ $^{42}_{20}Ca$

2. $^{239}Pu \rightarrow {}^{4}_{2}He + $ ___ $^{235}_{92}U$

3. $^{235}_{92}U \rightarrow $ ___ $^{4}_{2}He + {}^{231}_{90}Th$

4. $^{1}_{1}H + {}^{3}_{1}H \rightarrow $ ___ $^{4}_{2}He$

5. $^{6}_{3}Li + {}^{1}_{0}n \rightarrow {}^{4}_{2}He + $ ___ $^{3}_{1}H$

6. $^{27}_{13}Al + {}^{4}_{2}He \rightarrow {}^{30}_{15}P + $ ___ $^{1}_{0}N$

7. $^{9}_{4}Be + {}^{1}_{1}H \rightarrow $ ___ $^{6}_{3}Li + {}^{4}_{2}He$

8. $^{37}K \rightarrow {}^{0}_{-1}e + $ ___ $^{37}_{18}Ar$

9. $^{235}_{92}U $ ___ $ + {}^{1}_{0}n \rightarrow {}^{142}_{56}Ba + {}^{91}_{36}Kr + 3{}^{1}_{0}n$

10. $^{238}_{92}U + {}^{4}_{2}He \rightarrow $ ___ $^{241}_{94}Pu + {}^{1}_{0}n$

Half-Lives of Radioactive Isotopes

Solve each problem.

1. How much of a 100.0 g sample of ^{198}Au is left after 8.10 days if its half-life is 2.70 days?

   12.5 g

2. A 50.0 g sample of ^{16}N decays to 12.5 g in 14.4 seconds. What is its half-life?

   7.2 s

3. The half-life of ^{42}K is 12.4 hours. How much of a $75\overline{0}$ g sample is left after 62.0 hours?

   23.4 g

4. What is the half-life of ^{99}Tc if a $50\overline{0}$ g sample decays to 62.5 g in 639,000 years?

   2.13 × 10⁵ y

5. The half-life of ^{232}Th is 1.4×10^{10} years. If there are 25.0 g of the sample left after 2.8×10^{10} years, how many grams were in the original sample?

   $1\overline{0}0$ g

6. There are 5.0 g of ^{131}I left after 40.35 days. How many grams were in the original sample if its half-life is 8.07 days?

   160 g

Periodic Table Worksheet

Use a copy of the periodic table to answer each question.

1. Where are the most active metals located? **lower left**
2. Where are the most active nonmetals located? **upper right**
3. As you go from left to right across a period, the atomic size (decreases, increases). Why? **increased positive nuclear charge**
4. As you travel down a group, the atomic size (decreases, increases). Why? **additional principal energy levels**
5. A negative ion is (larger, smaller) than its parent atom.
6. A positive ion is (larger, smaller) than its parent atom.
7. As you go from left to right across a period, the first ionization energy generally (decreases, increases). Why? **increased positive nuclear charge**
8. As you go down a group, the first ionization energy generally (decreases, increases). Why? **outermost electron is farther away from nucleus, shielding effect of inner electrons**
9. Where is the highest electronegativity found? **upper right (F)**
10. Where is the lowest electronegativity found? **lower left (Fr)**
11. Elements of Group 1 are called **alkali metals**
12. Elements of Group 2 are called **alkaline earth metals**
13. Elements of Group 3–12 are called **transition elements**
14. As you go from left to right across the periodic table, the elements go from (metals, nonmetals) to (metals, nonmetals).
15. Group 17 elements are called **halogens**
16. The most active element in Group 17 is **fluorine**
17. Group 18 elements are called **noble gases**
18. What sublevels are filling across the Transition Elements? **d and f**
19. Elements within a group have a similar number of **valence electrons**
20. Elements across a series have the same number of **principal energy levels**
21. A colored ion generally indicates a **transition element**
22. As you go down a group, the elements generally become (more, less) metallic.
23. The majority of elements in the periodic table are (metals, nonmetals).
24. Elements in the periodic table are arranged according to their **atomic numbers**
25. An element with both metallic and nonmetallic properties is called a **semimetal or metalloid**

Answer Key

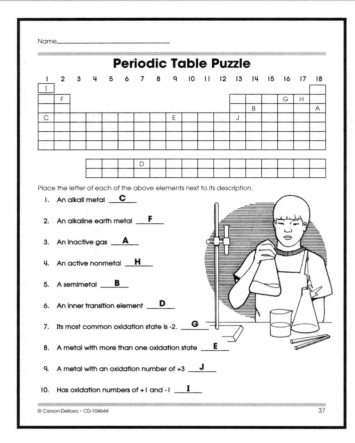

Periodic Table Puzzle

Place the letter of each of the above elements next to its description.

1. An alkali metal __C__

2. An alkaline earth metal __F__

3. An inactive gas __A__

4. An active nonmetal __H__

5. A semimetal __B__

6. An inner transition element __D__

7. Its most common oxidation state is -2. __G__

8. A metal with more than one oxidation state __E__

9. A metal with an oxidation number of +3 __J__

10. Has oxidation numbers of +1 and -1 __I__

© Carson-Dellosa • CD-104644   37

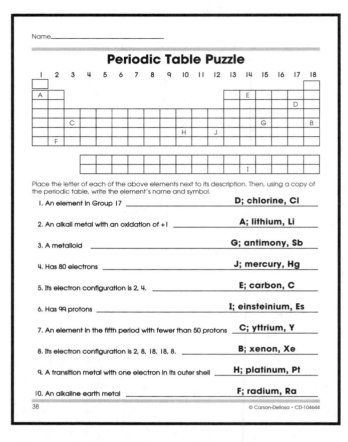

Periodic Table Puzzle

Place the letter of each of the above elements next to its description. Then, using a copy of the periodic table, write the element's name and symbol.

1. An element in Group 17 __D; chlorine, Cl__

2. An alkali metal with an oxidation of +1 __A; lithium, Li__

3. A metalloid __G; antimony, Sb__

4. Has 80 electrons __J; mercury, Hg__

5. Its electron configuration is 2, 4. __E; carbon, C__

6. Has 99 protons __I; einsteinium, Es__

7. An element in the fifth period with fewer than 50 protons __C; yttrium, Y__

8. Its electron configuration is 2, 8, 18, 18, 8. __B; xenon, Xe__

9. A transition metal with one electron in its outer shell __H; platinum, Pt__

10. An alkaline earth metal __F; radium, Ra__

38   © Carson-Dellosa • CD-104644

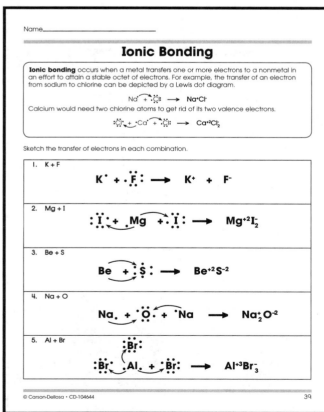

Ionic Bonding

Ionic bonding occurs when a metal transfers one or more electrons to a nonmetal in an effort to attain a stable octet of electrons. For example, the transfer of an electron from sodium to chlorine can be depicted by a Lewis dot diagram.

Calcium would need two chlorine atoms to get rid of its two valence electrons.

Sketch the transfer of electrons in each combination.

© Carson-Dellosa • CD-104644   39

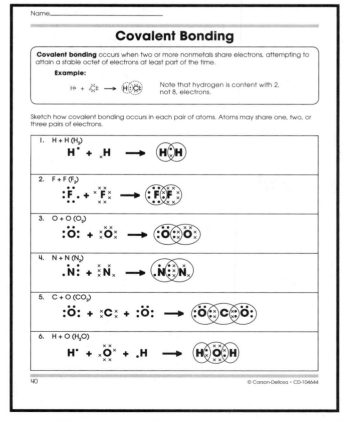

Covalent Bonding

Covalent bonding occurs when two or more nonmetals share electrons, attempting to attain a stable octet of electrons at least part of the time.

Example:

Sketch how covalent bonding occurs in each pair of atoms. Atoms may share one, two, or three pairs of electrons.

40   © Carson-Dellosa • CD-104644

Answer Key

Name_____

Types of Chemical Bonds

Identify each compound as *ionic* (metal + nonmetal), *covalent* (nonmetal + nonmetal), or *both* (compound containing a polyatomic ion).

1. $CaCl_2$ _____**ionic**_____
2. CO_2 _____**covalent**_____
3. H_2O _____**covalent**_____
4. $BaSO_4$ _____**both**_____
5. K_2O _____**ionic**_____
6. NaF _____**ionic**_____
7. Na_2CO_3 _____**both**_____
8. CH_4 _____**covalent**_____
9. SO_3 _____**covalent**_____
10. $LiBr$ _____**ionic**_____
11. MgO _____**ionic**_____
12. NH_4Cl _____**both**_____
13. HCl _____**covalent**_____
14. KI _____**ionic**_____
15. $NaOH$ _____**both**_____
16. NO_2 _____**covalent**_____
17. $AlPO_4$ _____**both**_____
18. $FeCl_3$ _____**ionic**_____
19. P_2O_5 _____**covalent**_____
20. N_2O_3 _____**covalent**_____

Name_____

Shapes of Molecules

Using VSEPR theory, name and sketch the shape of each molecule.

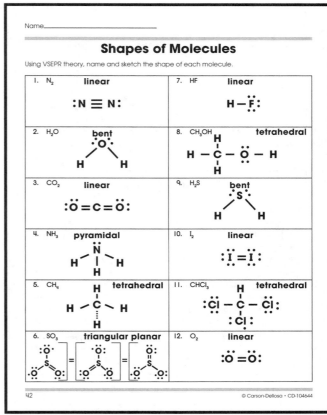

Name_____

Polarity of Molecules

Identify each molecule as *polar* or *nonpolar*.

1. N_2 **nonpolar**	7. HF **polar**
2. H_2O **polar**	8. CH_3OH **polar**
3. CO_2 **nonpolar**	9. H_2S **polar**
4. NH_3 **polar**	10. I_2 **nonpolar**
5. CH_4 **nonpolar**	11. $CHCl_3$ **polar**
6. SO_3 **nonpolar**	12. O_2 **nonpolar**

Name_____

Chemical Bonding Crossword

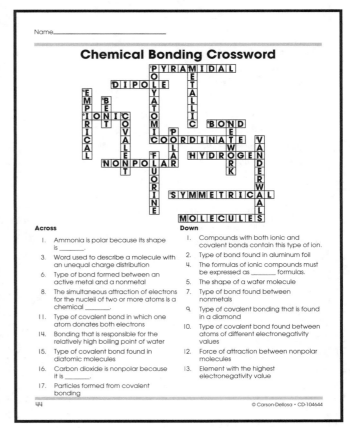

Across

1. Ammonia is polar because its shape is _____.
3. Word used to describe a molecule with an unequal charge distribution
6. Type of bond formed between an active metal and a nonmetal
8. The simultaneous attraction of electrons for the nuclei of two or more atoms is a chemical _____.
11. Type of covalent bond in which one atom donates both electrons
14. Bonding that is responsible for the relatively high boiling point of water
15. Type of covalent bond found in diatomic molecules
16. Carbon dioxide is nonpolar because it is _____.
17. Particles formed from covalent bonding

Down

1. Compounds with both ionic and covalent bonds contain this type of ion.
2. Type of bond found in aluminum foil
4. The formulas of ionic compounds must be expressed as _____ formulas.
5. The shape of a water molecule
7. Type of bond found between nonmetals
9. Type of covalent bonding that is found in a diamond
10. Type of covalent bond found between atoms of different electronegativity values
12. Force of attraction between nonpolar molecules
13. Element with the highest electronegativity value

Answer Key

Writing Formulas (Crisscross Method)

Write the formula of the compound produced from the listed ions.

	Cl^-	CO_3^{-2}	OH^-	SO_4^{-2}	PO_4^{-3}	NO_3^-
Na^+	$NaCl$	Na_2CO_3	$NaOH$	Na_2SO_4	Na_3PO_4	$NaNO_3$
NH_4^+	NH_4Cl	$(NH_4)_2CO_3$	NH_4OH	$(NH_4)_2SO_4$	$(NH_4)_3PO_4$	NH_4NO_3
K^+	KCl	K_2CO_3	KOH	K_2SO_4	K_3PO_4	KNO_3
Ca^{2+}	$CaCl_2$	$CaCO_3$	$Ca(OH)_2$	$CaSO_4$	$Ca_3(PO_4)_2$	$Ca(NO_3)_2$
Mg^{2+}	$MgCl_2$	$MgCO_3$	$Mg(OH)_2$	$MgSO_4$	$Mg_3(PO_4)_2$	$Mg(NO_3)_2$
Zn^{2+}	$ZnCl_2$	$ZnCO_3$	$Zn(OH)_2$	$ZnSO_4$	$Zn_3(PO_4)_2$	$Zn(NO_3)_2$
Fe^{3+}	$FeCl_3$	$Fe_2(CO_3)_3$	$Fe(OH)_3$	$Fe_2(SO_4)_3$	$FePO_4$	$Fe(NO_3)_3$
Al^{3+}	$AlCl_3$	$Al_2(CO_3)_3$	$Al(OH)_3$	$Al_2(SO_4)_3$	$AlPO_4$	$Al(NO_3)_3$
Co^{3+}	$CoCl_3$	$Co_2(CO_3)_3$	$Co(OH)_3$	$Co_2(SO_4)_3$	$CoPO_4$	$Co(NO_3)_3$
Fe^{2+}	$FeCl_2$	$FeCO_3$	$Fe(OH)_2$	$FeSO_4$	$Fe_3(PO_4)_2$	$Fe(NO_3)_2$
H^+	HCl	H_2CO_3	HOH or H_2O	H_2SO_4	H_3PO_4	HNO_3

© Carson-Dellosa • CD-104644   45

Naming Ionic Compounds

Name each compound using the Stock Naming System.

1. $CaCO_3$ — calcium carbonate
2. KCl — potassium chloride
3. $FeSO_4$ — iron(II) sulfate
4. $LiBr$ — lithium bromide
5. $MgCl_2$ — magnesium chloride
6. $FeCl_3$ — iron(III) chloride
7. $Zn_3(PO_4)_2$ — zinc phosphate
8. NH_4NO_3 — ammonium nitrate
9. $Al(OH)_3$ — aluminum hydroxide
10. $CuC_2H_3O_2$ — copper(I) acetate
11. $PbSO_3$ — lead(II) sulfite
12. $NaClO_3$ — sodium chlorate
13. CaC_2O_4 — calcium oxalate
14. Fe_2O_3 — iron(III) oxide
15. $(NH_4)_3PO_4$ — ammonium phosphate
16. $NaHSO_4$ — sodium hydrogen sulfate or sodium bisulfate
17. Hg_2Cl_2 — mercury(I) chloride
18. $Mg(NO_2)_2$ — magnesium nitrate
19. $CuSO_4$ — copper(II) sulfate
20. $NaHCO_3$ — sodium hydrogen carbonate or sodium bicarbonate
21. $NiBr_3$ — nickel(III) bromide
22. $Be(NO_3)_2$ — beryllium nitrate
23. $ZnSO_4$ — zinc sulfate
24. $AuCl_3$ — gold(III) chloride
25. $KMnO_4$ — potassium permanganate

46   © Carson-Dellosa • CD-104644

Naming Molecular Compounds

Name each covalent compound.

1. CO_2 — carbon dioxide
2. CO — carbon monoxide
3. SO_2 — sulfur dioxide
4. SO_3 — sulfur trioxide
5. N_2O — dinitrogen monoxide
6. NO — nitrogen monoxide
7. N_2O_3 — dinitrogen trioxide
8. NO_2 — nitrogen dioxide
9. N_2O_4 — dinitrogen tetroxide
10. N_2O_5 — dinitrogen pentoxide
11. PCl_3 — phosphorus trichloride
12. PCl_5 — phosphorus pentachloride
13. NH_3 — ammonia
14. SCl_6 — sulfur hexachloride
15. P_2O_5 — diphosphorus pentoxide
16. CCl_4 — carbon tetrachloride
17. SiO_2 — silicon dioxide
18. CS_2 — carbon disulfide
19. OF_2 — oxygen difluoride
20. PBr_3 — phosphorus tribromide

© Carson-Dellosa • CD-104644   47

Naming Acids

Name each acid.

1. HNO_3 — nitric acid
2. HCl — hydrochloric acid
3. H_2SO_4 — sulfuric acid
4. H_2SO_3 — sulfurous acid
5. $HC_2H_3O_2$ — acetic acid
6. HBr — hydrobromic acid
7. HNO_2 — nitrous acid
8. H_3PO_4 — phosphoric acid
9. H_2S — hydrosulfuric acid
10. H_2CO_3 — carbonic acid

Write the formula of each acid.

11. sulfuric acid — H_2SO_4
12. nitric acid — HNO_3
13. hydrochloric acid — HCl
14. acetic acid — $HC_2H_3O_2$
15. hydrofluoric acid — HF
16. phosphorous acid — H_3PO_3
17. carbonic acid — H_2CO_3
18. nitrous acid — HNO_2
19. phosphoric acid — H_3PO_4
20. hydrosulfuric acid — H_2S

48   © Carson-Dellosa • CD-104644

Answer Key

Writing Formulas from Names

Write the formula of each compound.

1.	ammonium phosphate	$(NH_4)PO_4$
2.	iron(II) oxide	FeO
3.	iron(III) oxide	Fe_2O_3
4.	carbon monoxide	CO
5.	calcium chloride	$CaCl_2$
6.	potassium nitrate	KNO_3
7.	magnesium hydroxide	$Mg(OH)_2$
8.	aluminum sulfate	$Al_2(SO_4)_3$
9.	copper(II) sulfate	$CuSO_4$
10.	lead(IV) chromate	$Pb(CrO_4)_2$
11.	diphosphorus pentoxide	P_2O_5
12.	potassium permanganate	$KMnO_4$
13.	sodium hydrogen carbonate	$NaHCO_3$
14.	zinc nitrate	$Zn(NO_3)_2$
15.	aluminum sulfite	$Al_2(SO_3)_3$

49

Gram Formula Mass

Determine the gram formula mass (the mass of one mole) of each compound.

1.	$KMnO_4$	158 g
2.	KCl	74.55 g
3.	Na_2SO_4	142 g
4.	$Ca(NO_3)_2$	164 g
5.	$Al_2(SO_4)_3$	342 g
6.	$(NH_4)_3PO_4$	149 g
7.	$CuSO_4 \cdot 5H_2O$	250 g
8.	$Mg_3(PO_4)_2$	262.86 g
9.	$Zn(C_2H_3O_2)_2 \cdot 2H_2O$	219 g
10.	$Zn_3(PO_4)_2 \cdot 4H_2O$	458 g
11.	H_2CO_3	62 g
12.	$Hg_2Cr_2O_7$	617 g
13.	$Ba(ClO_3)_2$	304 g
14.	$Fe_2(SO_3)_3$	352 g
15.	$NH_4C_2H_3O_2$	77 g

50

Moles and Mass

Determine the number of moles in each quantity.

1.	25 g of NaCl	0.43 mole
2.	125 g of H_2SO_4	1.28 moles
3.	100. g of $KMnO_4$	0.633 mole
4.	74 g of KCl	0.97 mole
5.	35 g of $CuSO_4 \cdot 5H_2O$	0.14 mole

Determine the number of grams in each quantity.

6.	2.5 mol of NaCl	145 g
7.	0.50 mol of H_2SO_4	49 g
8.	1.70 mol of $KMnO_4$	269 g
9.	0.25 mol of KCl	19 g
10.	3.2 mol of $CuSO_4 \cdot 5H_2O$	800 g

51

The Mole and Volume

For gases at STP (273 K and 1 atm pressure), one mole occupies a volume of 22.4 L. Identify the volume each quantity of gas will occupy at STP.

1.	1.00 mole of H_2	22.4 L
2.	3.20 moles of O_2	71.7 L
3.	0.750 mole of N_2	16.8 L
4.	1.75 moles of CO_2	39.2 L
5.	0.50 mole of NH_3	11.2 L
6.	5.0 g of H2	56 L
7.	100. g of O_2	70.0 L
8.	28.0 g of N_2	22.4 L
9.	60. g of CO_2	31 L
10.	10. g of NH_3	13 L

52

Answer Key

Name_____

The Mole and Avogadro's Number

One mole of a substance contains Avogadro's number (6.02×10^{23}) of molecules.

Identify how many molecules are in each quantity.

1. 2.0 mol	1.2×10^{24}
2. 1.5 mol	9.0×10^{23}
3. 0.75 mol	4.5×10^{23}
4. 15 mol	9.0×10^{24}
5. 0.35 mol	2.1×10^{23}

Identify how many moles are in the number of molecules listed.

6. 6.02×10^{23}	1.00 moles
7. 1.204×10^{24}	2.00 moles
8. 1.5×10^{20}	2.5×10^{-5} or 0.000025 mole
9. 3.4×10^{26}	560 moles
10. 7.5×10^{19}	1.2×10^{-4} or 0.00012 mole

Name_____

Mixed Mole Problems

Solve each problem.

1. How many grams are there in 1.5×10^{25} molecules of CO_2?

1.1×10^3 g

2. What volume would the CO_2 in problem 1 occupy at STP?

5.6×10^2 L

3. A sample of NH_3 gas occupies 75.0 liters at STP. How many molecules is this?

2.02×10^{24} molecules

4. What is the mass of the sample of NH_3 in problem 3?

56.9 g

5. How many atoms are there in 1.3×10^{22} molecules of NO_2?

3.9×10^{22} atoms

6. A 5.0 g sample of O_2 is in a container at STP. What volume is the container?

3.5 L

7. How many molecules of O_2 are in the container in problem 6? How many atoms of oxygen?

9.4×10^{22} molecules, 1.9×10^{23} atoms

Name_____

Percentage Composition

Determine the percentage composition of each compound.

1. $KMnO_4$
   K = __24.7%__
   Mn = __34.8%__
   O = __40.5%__

2. HCl
   H = __2.8%__
   Cl = __97.2%__

3. $Mg(NO_3)_2$
   Mg = __16.2%__
   N = __18.9%__
   O = __64.9%__

4. $(NH_4)PO_4$
   N = __12.4%__
   H = __3.6%__
   P = __27.4%__
   O = __56.6%__

5. $Al_2(SO_4)_3$
   Al = __15.8%__
   S = __28.1%__
   O = __56.1%__

Solve each problem.

6. How many grams of oxygen can be produced from the decomposition of 100. g of $KClO_3$? __39.3 g O__

7. How much iron can be recovered from 25.0 g of Fe_2O_3? __17.5 g Fe__

8. How much silver can be produced from 125 g of Ag_2S? __109 g Ag__

Name_____

Determining Empirical Formulas

Identify the empirical formula (lowest whole number ratio) of each compound.

1. 75% carbon, 25% hydrogen

CH_5

2. 52.7% potassium, 47.3% chlorine

KCl

3. 22.1% aluminum, 25.4% phosphorus, 52.5% oxygen

$AlPO_4$

4. 13% magnesium, 87% bromine

$MgBr_2$

5. 32.4% sodium, 22.5% sulfur, 45.1% oxygen

Na_2SO_4

6. 25.3% copper, 12.9% sulfur, 25.7% oxygen, 36.1% water

$CuSO_4(H_2O)_5$

Answer Key

Name_____

Determining Molecular Formulas
(True Formulas)

Solve each problem.

1. The empirical formula of a compound is NO_2. Its molecular mass is 92 g/mol. What is its molecular formula?

$$N_2O_4$$

2. The empirical formula of a compound is CH_2. Its molecular mass is 70 g/mol. What is its molecular formula?

$$C_5H_{10}$$

3. A compound is found to be 40.0% carbon, 6.7% hydrogen and 53.5% oxygen. Its molecular mass is 60. g/mol. What is its molecular formula?

$$C_2H_4O_2 \quad (CH_3COOH)$$

4. A compound is 64.9% carbon, 13.5% hydrogen, and 21.6% oxygen. Its molecular mass is 74 g/mol. What is its molecular formula?

$$C_4H_{10}O \quad (C_4H_9OH)$$

5. A compound is 54.5% carbon, 9.1% hydrogen, and 36.4% oxygen. Its molecular mass is 88 g/mol. What is its molecular formula?

$$C_4H_8O_2 \quad (C_3H_7COOH)$$

Name_____

Composition of Hydrates

A **hydrate** is an ionic compound with water molecules loosely bonded to its crystal structure. The water is in a specific ratio to each formula unit of the salt. For example, the formula $CuSO_4 \cdot 5H_2O$ indicates that there are five water molecules for every one formula unit of $CuSO_4$.

Solve each problem.

1. What percentage of water is found in $CuSO_4 \cdot 5H_2O$?

35%

2. What percentage of water is found in $Na_2S \cdot 9H_2O$?

67.5%

3. A 5.0 g sample of a hydrate of $BaCl_2$ was heated, and only 4.3 g of the anhydrous salt remained. What percentage of water was in the hydrate?

14%

4. A 2.5 g sample of a hydrate of $Ca(NO_3)_2$ was heated, and only 1.7 g of the anhydrous salt remained. What percentage of water was in the hydrate?

32%

5. A 3.0 g sample of $Na_2CO_3 \cdot H_2O$ is heated to constant mass. How much anhydrous salt remains?

2.6 g

6. A 5.0 g sample of $Cu(NO_3)_2 \cdot nH_2O$ is heated to constant mass. How much anhydrous salt remains?

3.9 g

Name_____

Balancing Chemical Equations

Rewrite and balance each equation.

1. $N_2 + H_2 \rightarrow NH_3$ $N_2 + 3H_2 \rightarrow 2NH_3$

2. $KClO_3 \rightarrow KCl + O_2$ $2KClO_3 \rightarrow 2KCl + 3O_2$

3. $NaCl + F_2 \rightarrow NaF + Cl_2$ $2NaCl + F \rightarrow 2NaF + Cl_2$

4. $H_2 + O_2 \rightarrow H_2O$ $2H_2 + O_2 \rightarrow 2H_2O$

5. $AgNO_3 + MgCl_2 \rightarrow AgCl + Mg(NO_3)_2$ $2AgNO_3 + MgCl_2 \rightarrow 2AgCl + Mg(NO_3)_2$

6. $AlBr_3 + K_2SO \rightarrow KBr + Al_2(SO_4)_3$ $2AlBr_3 + 3K_2SO_4 \rightarrow 6KBr + Al_2(SO_4)_3$

7. $CH_4 + O_2 \rightarrow CO_2 + H_2O$ $CH_4 + 2O_2 \rightarrow CO_2 + 2H_2O$

8. $C_3H_6 + O_2 \rightarrow CO_2 + H_2O$ $C_3H_6 + 5O_2 \rightarrow 3CO_2 + 4H_2O$

9. $C_6H_{18} + O_2 \rightarrow CO_2 + H_2O$ $2C_6H_{18} + 25O_2 \rightarrow 16CO_2 + 18H_2O$

10. $FeCl_3 + NaOH \rightarrow Fe(OH)_3 + NaCl$ $FeCl_3 + 3NaOH \rightarrow Fe(OH)_3 + 3NaCl$

11. $P + O_2 \rightarrow P_2O_5$ $4P + 5O_2 \rightarrow 2P_2O_5$

12. $Na + H_2O \rightarrow NaOH + H_2$ $2Na + 2H_2O \rightarrow 2NaOH + H_2$

13. $Ag_2O \rightarrow Ag + O_2$ $2Ag_2O \rightarrow 4Ag + O_2$

14. $S_8 + O_2 \rightarrow SO_3$ $S_8 + 12O_2 \rightarrow 8SO_3$

15. $CO_2 + H_2O \rightarrow C_6H_{12}O_6 + O_2$ $6CO_2 + 6H_2O \rightarrow C_6H_{12}O_6 + 6O_2$

16. $K + MgBr_2 \rightarrow KBr + Mg$ $2K + MgBr_2 \rightarrow 2KBr + Mg$

17. $HCl + CaCO_3 \rightarrow CaCl_2 + H_2O + CO_2$ $2HCl + CaCO_3 \rightarrow CaCl_2 + H_2O + CO_2$

Name_____

Word Equations

Rewrite each word equation as a chemical equation. Then, balance the equation.

1. zinc + lead(II) nitrate yield zinc nitrate + lead

$$Zn + Pb(NO_3)_2 \rightarrow Zn(NO_3)_2 + Pb$$

2. aluminum bromide + chlorine yield aluminum chloride + bromine

$$2AlBr_3 + 3Cl_2 \rightarrow 2AlCl_3 + 3Br_2$$

3. sodium phosphate + calcium chloride yield calcium + sodium chloride

$$2Na_3PO_4 + 3CaCl_2 \rightarrow Ca3(PO_4)_2 + 6NaCl$$

4. potassium chlorate, when heated, yields potassium chloride + oxygen gas

$$2KClO_3 \rightarrow 2KCl + 3O_2(g)$$

5. aluminum + hydrochloric acid yield aluminum chloride + hydrogen gas

$$2Al + 6HCl \rightarrow 2AlCl_3 + 3H_2(g)$$

6. calcium hydroxide + phosphoric acid yield calcium phosphate + water

$$3Ca(OH)_2 + 2H_3PO_4 \rightarrow Ca_3(PO_4)_2 + 6H_2O$$

7. copper + sulfuric acid yield copper(II) sulfate + water + sulfur dioxide

$$Cu + 2H_2SO_4 \rightarrow CuSO_4 + 2H_2O + SO_2$$

8. hydrogen + nitrogen monoxide yield water + nitrogen

$$2H_2 + 2NO \rightarrow 2H_2O + N_2$$

Answer Key

Name_____

Classification of Chemical Reactions

Identify each reaction as *synthesis, decomposition, cationic* or *anionic single replacement*, or *double replacement*.

1. $2H_2 + O_2 \rightarrow 2H_2O$

    synthesis

2. $2H_2O \rightarrow 2H_2 + O_2$

    decomposition

3. $Zn + H_2SO_4 \rightarrow ZnSO_4 + H_2$

    cationic single replacement

4. $2CO + O_2 \rightarrow 2CO_2$

    synthesis

5. $2HgO \rightarrow 2Hg + O_2$

    decomposition

6. $2KBr + Cl_2 \rightarrow 2KCl + Br_2$

    anionic single replacement

7. $CaO + H_2O \rightarrow Ca(OH)_2$

    synthesis

8. $AgNO + NaCl \rightarrow AgCl + NaNO_3$

    double replacement

9. $2H_2O_2 \rightarrow 2H_2O + O_2$

    decomposition

10. $Ca(OH)_2 + H_2SO_4 \rightarrow CaSO_4 + 2H_2O$

    double replacement

Name_____

Predicting Products of Chemical Reactions

Predict the product in each reaction. Then, write the balanced equation and classify the reaction. **Predictions will vary.**

1. magnesium bromide + chlorine   **anionic single replacement**

    $MgBr_2 + Cl_2 \rightarrow MgCl_2Br_2$

2. aluminum + iron(III) oxide   **cationic single replacement**

    $2Al + Fe_2O_3 \rightarrow 2Fe + Al_2O_3$

3. silver nitrate + zinc chloride   **double replacement**

    $2AgNO_3 + ZnCl_2 \rightarrow 2AgCl + Zn(NO_3)_2$

4. hydrogen peroxide (catalyzed by manganese dioxide)   **decomposition**

    $2H_2O_2 \xrightarrow{MnO_2} 2H_2O + O_2$

5. zinc + hydrochloric acid   **cationic single replacement**

    $Zn + 2HCl \rightarrow ZnCl_2 + H_2$

6. sulfuric acid + sodium hydroxide   **double replacement (neutralization)**

    $H_2SO_4 + 2NaOH \rightarrow Na_2SO_4 + 2H_2O$

7. sodium + hydrogen   **synthesis**

    $2Na + H_2 \rightarrow 2NaOH$

8. acetic acid + copper   **none**

    $CH_3COOH \text{ (or } HC_2H_3O_2) + Cu \rightarrow \text{ no reaction}$

Name_____

Stoichiometry: Mole-Mole Problems

Solve each problem.

1. $N_2 + 3H_2 \rightarrow 2NH_3$
   How many moles of hydrogen are needed to completely react with two moles of nitrogen?

    6 moles

2. $2KClO_3 \rightarrow 2KCl + 3O_2$
   How many moles of oxygen are produced by the decomposition of six moles of potassium chlorate?

    9 moles

3. $Zn + 2HCl \rightarrow ZnCl_2 + H_2$
   How many moles of hydrogen are produced from the reaction of three moles of zinc with an excess of hydrochloric acid?

    3 moles

4. $C_3H_8 + 5O_2 \rightarrow 3CO_2 + 4H_2O$
   How many moles of oxygen are necessary to react completely with four moles of propane (C_3H_8)?

    20 moles

5. $K_3PO_4 + Al(NO_3)_3 \rightarrow 3KNO_3 + AlPO_4$
   How many moles of potassium nitrate (KNO_3) are produced when six moles of potassium phosphate (KPO_4) react with two moles of aluminum nitrate($Al(NO_3)_3$)?

    6 moles

Name_____

Stoichiometry: Volume-Volume Problems

Solve each problem.

1. $N_2 + 3H_2 \rightarrow 2NH_3$
   What volume of hydrogen is necessary to react with five liters of nitrogen to produce ammonia? (Assume constant temperature and pressure.)

    15 L

2. What volume of ammonia is produced in the reaction in problem 1?

    10 L

3. $C_3H_8 + 5O_2 \rightarrow 3CO_2 + 4H_2O$
   If 20 liters of oxygen are consumed in the above reaction, how many liters of carbon dioxide are produced?

    12 L

4. $2H_2O \rightarrow 2H + O_2$
   If 30 mL of hydrogen are produced in the above reaction, how many milliliters of oxygen are produced?

    15 mL

5. $2CO + O_2 \rightarrow 2CO_2$
   How many liters of carbon dioxide are produced if 75 liters of carbon monoxide are burned in oxygen? How many liters of oxygen are necessary?

    75 L CO₂
    37.5 L O₂

Answer Key

Stoichiometry: Mass-Mass Problems

Solve each problem.

1. $2KClO_3 \rightarrow 2KCl + 3O_2$
   How many grams of potassium chloride are produced if 25 g of potassium chlorate decompose?

   15 g

2. $N_2 + 3H_2 \rightarrow 2NH_3$
   How many grams of hydrogen are necessary to react completely with 50.0 g of nitrogen in the above reaction?

   10.7 g

3. How many grams of ammonia are produced in the reaction in problem 2?

   60.7 g

4. $2AgNO_3 + BaCl \rightarrow 2AgCl + Ba(NO_3)$
   How many grams of silver chloride are produced from 5.0 g of silver nitrate reacting with an excess of barium chloride?

   4.2 g

5. How much barium chloride is necessary to react with the silver nitrate in problem 4?

   3.1 g

Stoichiometry: Mixed Problems

Solve each problem.

1. $N_2 + 3H_2 \rightarrow 2NH_3$
   What volume of NH_3 at STP is produced if 25.0 g of N_2 is reacted with an excess of H_2?

   40.0 L

2. $2KClO_3 \rightarrow 2KCl + 3O_2$
   If 5.0 g of $KClO_3$ are decomposed, what volume of O_2 is produced at STP?

   1.4 L

3. How many grams of KCl are produced in problem 2?

   3.0 g

4. $Zn + 2HCl \rightarrow ZnCl_2 + H_2$
   What volume of hydrogen at STP is produced when 2.5 g of zinc react with an excess of hydrochloric acid?

   0.86 L

5. $H_2SO_4 + 2NaOH \rightarrow H_2O + Na_2SO_4$
   How many molecules of water are produced if 2.0 g of sodium sulfate are produced in the above reaction?

   8.5×10^{21} molecules

6. $2AlCl_3 \rightarrow 2Al + 3Cl_2$
   If 10.0 g of aluminum chloride are decomposed, how many molecules of Cl_2 are produced?

   6.77×10^{22} molecules

Stoichiometry: Limiting Reagent

Solve each problem.

1. $N_2 + 3H_2 \rightarrow 2NH_3$
   How many grams of NH_3 can be produced from the reaction of 28 g of N_2 and 25 g of H_2?

   34 g

2. How much of the excess reagent in problem 1 is left over?

   19 g

3. $Mg + 2HCl \rightarrow MgCl_2 + H_2$
   What volume of hydrogen at STP is produced from the reaction of 50.0 g of Mg and the equivalent of 75 g of HCl?

   23.0 L

4. How much of the excess reagent in problem 3 is left over?

   24.3 g

5. $3AgNO_3 + Na_3PO_4 \rightarrow Ag_3PO_4 + 3NaNO_3$
   Silver nitrate and sodium phosphate are reacted in equal amounts of 200. g each. How many grams of silver phosphate are produced?

   165 g

6. How much of the excess reagent in problem 5 is left over?

   136 g

Solubility Curves

Answer each question based on the solubility curve shown.

1. Which salt is least soluble in water at 20°C? **$KClO_3$**

2. How many grams of potassium chloride can be dissolved in 200 g of water at 80°C? **100 g**

3. At 40°C, how much potassium nitrate can be dissolved in 300 g of water? **126 g**

4. Which salt shows the least change in solubility from 0°C to 100°C? **NaCl**

5. At 30°C, 85 g of sodium nitrate are dissolved in 100 g of water. Is this solution *saturated, unsaturated,* or *supersaturated*? **unsaturated**

6. A saturated solution of potassium chlorate is formed from 100 g of water. If the saturated solution is cooled from 80°C to 50°C, how many grams of precipitate are formed? **7 g**

7. What compound shows a decrease in solubility from 0°C to 100°C? **NH_3**

8. Which salt is most soluble at 10°C? **KI**

9. Which salt is least soluble at 50°C? **$KClO_3$**

10. Which salt is least soluble at 90°C? **NH_3**

Answer Key

Name_____

Molarity (M)

$$\text{Molarity} = \frac{\text{moles of solute}}{\text{liter of solution}}$$

Solve each problem.

1. What is the molarity of a solution in which 58 g of NaCl are dissolved in 1.0 L of solution?

   1.0 M

2. What is the molarity of a solution in which 10.0 g of $AgNO_3$ are dissolved in 500. mL of solution?

   0.118 M

3. How many grams of KNO_3 should be used to prepare 2.00 L of a 0.500 M solution?

   101 g

4. To what volume should 5.0 g of KCl be diluted in order to prepare a 0.25 M solution?

   270 mL

5. How many grams of $CuSO_4 \cdot 5H_2O$ are needed to prepare 100. mL of a 0.10 M solution?

   2.5 g

Name_____

Molarity by Dilution

Acids are usually acquired from chemical supply houses in concentrated form. These acids are diluted to the desired concentration by adding water. Since moles of acid before dilution equal moles of acid after dilution, and moles of acid = $M \times V$, then $M_1 \times V_1 = M_2 \times V_2$.

Solve each problem.

1. How much concentrated 18 M sulfuric acid is needed to prepare 250 mL of a 6.0 M solution?

   83 mL

2. How much concentrated 12 M hydrochloric acid is needed to prepare 100 mL of a 2.0 M solution?

   17 mL

3. To what volume should 25 mL of 15 M nitric acid be diluted to prepare a 3.0 M solution?

   125 mL

4. How much water should be added to 50. mL of 12 M hydrochloric acid to produce a 4.0 M solution?

   100 mL (150 mL total solution)

5. How much water should be added to 100. mL of 18 M sulfuric acid to prepare a 1.5 M solution?

   1.1 L (1.2 L or 1,200 mL total solution)

Name_____

Molality (m)

$$\text{Molality} = \frac{\text{moles of solute}}{\text{kg of solvent}}$$

Solve each problem.

1. What is the molality of a solution in which 3.0 moles of NaCl are dissolved in 1.5 kg of water?

   2.0 m

2. What is the molality of a solution in which 25 g of NaCl are dissolved in 2.0 kg of water?

   0.22 m

3. What is the molality of a solution in which 15 g of I_2 are dissolved in 500. g of alcohol?

   0.12 m

4. How many grams of I_2 should be added to 750 g of CCl_4 to prepare a 0.020 m solution?

   3.8 g

5. How much water should be added to 5.00 g of KCl to prepare a 0.500 m solution?

   135 g

Name_____

Normality (N)

normality = molarity × total positive oxidation number of solute

Example: What is the normality of 3.0 M of H_2SO_4?

Since the total positive oxidation number of H_2SO_4 is +2 (2 H+), N = 6.0.

Solve each problem.

1. What is the normality of a 2.0 M NaOH solution?

   2.0 N

2. What is the normality of a 2.0 M H_3PO_4 solution?

   6.0 N

3. A solution of H_2SO_4 is 3.0 N. What is its molarity?

   1.5 M

4. What is the normality of a solution in which 2.0 g of $Ca(OH)_2$ is dissolved in 1.0 L of solution?

   0.054 N

5. How much $AlCl_3$ should be dissolved in 2.00 L of solution to produce a 0.150 N solution?

   13.3 g

Answer Key

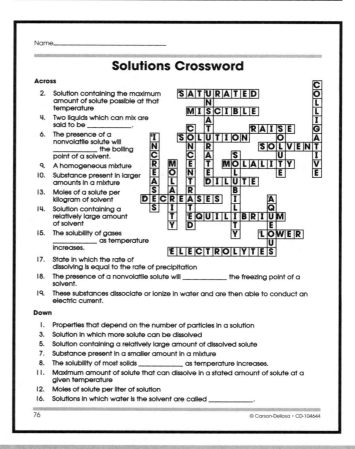
© Carson-Dellosa • CD-104644   121

Answer Key

Potential Energy Diagram

Name_____

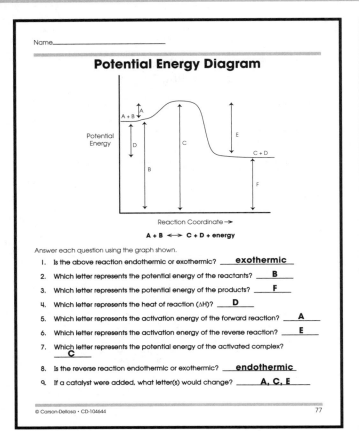

Potential Energy

Reaction Coordinate →

$A + B \longleftrightarrow C + D + energy$

Answer each question using the graph shown.

1. Is the above reaction endothermic or exothermic? **exothermic**

2. Which letter represents the potential energy of the reactants? **B**

3. Which letter represents the potential energy of the products? **F**

4. Which letter represents the heat of reaction (ΔH)? **D**

5. Which letter represents the activation energy of the forward reaction? **A**

6. Which letter represents the activation energy of the reverse reaction? **E**

7. Which letter represents the potential energy of the activated complex? **C**

8. Is the reverse reaction endothermic or exothermic? **endothermic**

9. If a catalyst were added, what letter(s) would change? **A, C, E**

77

Entropy

Name_____

Entropy is the degree of randomness in a substance. The symbol for change in entropy is ΔS.

Solids are very ordered and have low entropy. Liquids and aqueous ions have more entropy because they move about more freely. Gases have an even larger amount of entropy. According to the Second Law of Thermodynamics, nature is always proceeding to a state of higher entropy.

Determine whether each reaction shows an *increase* or *decrease* in entropy.

1. $2KClO_3(s) \rightarrow 2KCl(s) + 3O_2(g)$ — **increase**

2. $H_2O(l) \rightarrow H_2O(s)$ — **decrease**

3. $N_2(g) + 3H_2(g) \rightarrow 2NH_3(g)$ — **decrease**

4. $NaCl(s) \rightarrow Na^+(aq) + Cl^-(aq)$ — **increase**

5. $KCl(s) \rightarrow KCl(l)$ — **increase**

6. $CO_2(s) \rightarrow CO_2(g)$ — **increase**

7. $H^+(aq) + C_2H_3O_2^-(aq) \rightarrow HC_2H_3O_2(l)$ — **decrease**

8. $C(s) + O_2(g) \rightarrow CO_2(g)$ — **increase**

9. $H_2(g) + Cl_2(g) \rightarrow 2HCl(g)$ — **no change**

10. $Ag^+(aq) + Cl^-(aq) \rightarrow AgCl(s)$ — **increase**

11. $2N_2O_5(g) \rightarrow 4NO_2(g) + O_2(g)$ — **increase**

12. $2Al(s) + 3I_2(s) \rightarrow 2AlI_3(s)$ — **decrease**

13. $H^+(aq) + OH^-(aq) \rightarrow H_2O(l)$ — **decrease**

14. $2NO(g) \rightarrow N_2(g) + O_2(g)$ — **no change**

15. $H_2O(g) \rightarrow H_2O(l)$ — **decrease**

78

Gibbs Free Energy

Name_____

For a reaction to be spontaneous, the sign of ΔG (Gibbs free energy) must be negative. The mathematical formula for this value is:

$$\Delta G = \Delta H - T\Delta S$$

where ΔH = change in enthalpy or heat of reaction

T = temperature in Kelvin

ΔS = change in entropy or randomness

Complete the table for the sign of ΔG: +, −, or *undetermined*. When conditions allow for an undetermined sign of ΔG, temperature will decide spontaneity.

ΔH	ΔS	ΔG
−	+	−
+	−	+
−	−	undetermined
+	+	undetermined

Answer each question.

1. The conditions in which ΔG is always negative are when ΔH is **negative** and ΔS is **positive**.

2. The conditions in which ΔG is always positive are when ΔH is **positive** and ΔS is **negative**.

3. When the situation is indeterminate, a low temperature favors the (entropy, ⟨enthalpy⟩) factor and a high temperature favors the (⟨entropy⟩, enthalpy) factor.

Answer problems 4–6 with *always*, *sometimes*, or *never*.

4. The reaction: $Na(OH)_s \rightarrow Na^+(aq) + OH^-(aq) + energy$ will **always** be spontaneous.

5. The reaction: $energy + 2H_2(g) + O_2(g) + 2H_2O(l)$ will **never** be spontaneous.

6. The reaction: $energy + H_2O(s) \rightarrow H_2O(l)$ will **sometimes** be spontaneous.

7. What is the value of ΔG if ΔH = 32.0 kJ, ΔS = +25.0 kJ/K and T = 293 K? **-7,293 k J**

8. Is the reaction in problem 7 spontaneous? **yes**

9. What is the value of ΔG if ΔH = +12.0 kJ, ΔS = 5.00 kJ/K and T = 290. K? **+1,460 k J**

10. Is the reaction in problem 9 spontaneous? **no**

79

Equilibrium Constant (K)

Name_____

Write the expression for the equilibrium constant (K) for each reaction.

1. $N_2(g) + 3H_2(g) \longleftrightarrow 2NH_3(g)$

$$K = \frac{(NH_3)^2}{(N_2)(H_2)^3}$$

2. $2KClO_3(s) \longleftrightarrow 2KCl(s) + 3O_2(g)$

$$K = (O_2)^3$$

3. $H_2O(l) \longleftrightarrow H^+(aq) + OH^-(aq)$

$$K = (H^+)(OH^-)$$

4. $2CO(g) + O_2(g) \longleftrightarrow 2CO_2(g)$

$$K = \frac{(CO_2)^2}{(CO)^2(O_2)}$$

5. $Li_2CO_3(s) \rightarrow 2Li^+(aq) + CO_3^{-2}(aq)$

$$K = (Li^+)^2(CO_3^{-2})$$

80

Answer Key

Name_____

Calculations Using the Equilibrium Constant

Using the equilibrium constant expressions you determined on page 80, calculate the value of K when:

1. $(NH_3) = 0.0100$ M, $(N_2) = 0.0200$ M, $(H_2) = 0.0200$ M

$$\frac{(0.01)^2}{(0.02)(0.02)^3} = 625$$

2. $(O_2) = 0.0500$ M

$$(0.05)^3 = 1.25 \times 10^{-4}$$

3. $(H^+) = 1 \times 10^{-8}$ M, $(OH^-) = 1 \times 10^{-6}$ M

$$(1 \times 10^{-8})(1 \times 10^{-6}) = 1 \times 10^{-14}$$

4. $(CO) = 2.0$ M, $(O_2) = 1.5$ M, $(CO_2) = 3.0$ M

$$\frac{(3.0)^2}{(2.0)^2(1.5)} = 1.5$$

5. $(Li^+) = 0.2$ M, $(CO_3^{-2}) = 0.1$ M

$$(0.2)^2(0.1) = 4 \times 10^{-3}$$

81

Name_____

Le Chatelier's Principle

Le Chatelier's principle states that when a system at equilibrium is subjected to a stress, the system will shift its equilibrium point in order to relieve the stress.

Complete the chart by writing *left*, *right*, or *none* for equilibrium shift. Then, write *decreases*, *increases*, or *remains the same* for the concentrations of reactants and products, and for the value of K. The first one has been done for you.

$$N_2(g) + 3H_2(g) \longleftrightarrow 2NH_3(g) + 22.0 \text{ kcal}$$

Stress	Equilibrium Shift	(N_2)	(H_2)	(NH_3)	K
1. Add N_2	right	———	decreases	increases	remains the same
2. Add H_2	right	decreases	———	increases	same
3. Add NH_3	left	increases	increases	———	same
4. Remove N_2	left	———	increases	decreases	same
5. Remove H_2	left	increases	———	decreases	same
6. Remove NH_3	right	decreases	decreases	———	same
7. Increase temperature	left	increases	increases	decreases	increases
8. Decrease temperature	right	decreases	decreases	increases	decreases
9. Increase pressure	right	decreases	decreases	increases	same
10. Decrease pressure	left	increases	increases	decreases	same

82

Name_____

Le Chatelier's Principle (Cont.)

$$12.6 \text{ kcal} + H_2(g) + I_2(g) \longleftrightarrow 2HI(g)$$

Stress	Equilibrium Shift	(H_2)	(I_2)	(HI)	K
11. Add H_2	right	———	decreases	increases	remains the same
12. Add I_2	right	decreases	———	increases	same
13. Add HI	left	increases	increases	———	same
14. Remove H_2	left	———	increases	decreases	same
15. Remove I_2	left	increases	———	decreases	same
16. Remove HI	right	decreases	decreases	———	same
17. Increase temperature	right	decreases	decreases	increases	increases
18. Decrease temperature	left	increases	increases	decreases	decreases
19. Increase pressure	none	same	same	same	same
20. Decrease pressure	none	same	same	same	same

$$NAOH(s) \longleftrightarrow Na^+(aq) + OH^-(aq) + 10.6 \text{ kcal}$$

Stress	Equilibrium Shift	Amount NaOH(s)	(Na^+)	(OH^-)	K
21. Add NaOH(s)	none		same	same	same
22. Add NaCl (Adds Na^+)	left	increases	———	decreases	same
23. Add KOH (Adds OH^-)	left	increases	decreases	———	same
24. Add H^+ (Removes OH^-)	right	decreases	increases	———	same
25. Increase temperature	left	increases	decreases	decreases	decreases
26. Decrease temperature	right	decreases	increases	increases	increases
27. Increase pressure	none	same	same	same	same
28. Decrease pressure	none	same	same	same	same

83

Name_____

Bronsted–Lowry Acids and Bases

According to **Bronsted-Lowry theory**, an acid is a proton (H^+) donor and a base is a proton acceptor.

Example: $HCl + OH^- \rightarrow Cl^- + H_2O$

The HCl acts as an acid, and the OH^- acts as a base. This reaction is reversible in that the H_2O can give back the proton to the Cl^-.

Label the Bronsted-Lowry acids and bases in each reaction and show the direction of proton transfer.

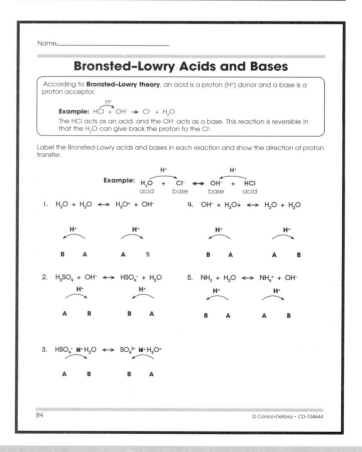

Example: $H_2O + Cl^- \longleftrightarrow OH^- + HCl$
acid   base   base   acid

1. $H_2O + H_2O \longleftrightarrow H_3O^+ + OH^-$ 4. $OH^- + H_3O^+ \longleftrightarrow H_2O + H_2O$

B A   A B   B A   A B

2. $H_2SO_4 + OH^- \longleftrightarrow HSO_4^- + H_2O$ 5. $NH_3 + H_2O \longleftrightarrow NH_4^+ + OH^-$

A B   B A   B A   A B

3. $HSO_4^- + H_2O \longleftrightarrow SO_4^{2-} + H_3O^+$

A B   B A

84

Answer Key

Conjugate Acid-Base Pairs

In the exercise on page 84, it was shown that after an acid has given up its proton, it is capable of getting the proton back and acting as a base. **Conjugate base** is what is left after an acid gives up a proton. The stronger the acid, the weaker the conjugate base. The weaker the acid, the stronger the conjugate base.

Complete the chart.

Conjugate Pairs

	Acid	Base	Equation
1.	H_2SO_4	HSO_4^-	$H_2SO_4 \longleftrightarrow H^+ + HSO_4^-$
2.	H_3PO_4	$H_2PO_4^-$	$H_3PO_4 \longrightarrow H^+ + H_2PO_4^-$
3.	HF	F^-	$HF \longrightarrow H^+ + F^-$
4.	HNO_3	NO_3^-	$HNO_3 \longleftrightarrow H^+ + NO_3^-$
5.	$H_2PO_4^-$	HPO_4^{-2}	$H_2PO_4^- \longleftrightarrow H^+$
6.	H_2O	OH^-	$H_2O \longleftrightarrow H^+ + OH^-$
7.	HSO_4^-	SO_4^{2-}	$HSO_4^- \longleftrightarrow H^+ + SO_4^{-3}$
8.	HPO_4^{-2}	PO_4^{-3}	$HPO_4^{-2} \longleftrightarrow H^+ + PO_4^{-3}$
9.	NH_4^+	NH_3	$NH_4^+ \longleftrightarrow H^+ + NH_3$
10.	H_3O^+	H_2O	$H_3O \longleftrightarrow H^+ + H_2O$

11. Which is a stronger base, HSO_4^- or $H_2PO_4^-$? ___$H_2PO_4^-$___

12. Which is a weaker base, Cl^- or NO_2^- ? ___Cl^-___

---

pH and pOH

The pH of a solution indicates how acidic or basic that solution is. A pH of less than 7 is acidic; a pH of 7 is neutral; and a pH greater than 7 is basic.

Since $(H^+)(OH^-) = 10^{-14}$ at 25°C, if (H^+) is known, the (OH^-) can be calculated and vice versa.

$$pH = -\log(H^+) \quad \text{So if } (H^+) = 10^{-6} M, pH = 6.$$
$$pOH = -\log(OH^-) \quad \text{So if } (OH^-) = 10^{-8} M, pOH = 8.$$
$$\text{Together, } pH + pOH = 14.$$

Complete the chart.

	(H^+)	pH	(OH^-)	pOH	Acidic or Basic
1.	10^{-5} M	5	10^{-9} M	9	acidic
2.	10^{-7} M	7	10^{-7} M	7	neutural
3.	10^{-10} M	10	10^{-4} M	14	basic
4.	10^{-2} M	2	10^{-12} M	12	acidic
5.	10^{-3} M	3	10^{-11} M	11	acidic
6.	10^{-12} M	12	10^{-2} M	2	basic
7.	10^{-9} M	9	10^{-5} M	5	basic
8.	10^{-11} M	11	10^{-3} M	3	basic
9.	10^{-1} M	1	10^{-13} M	13	acidic
10.	10^{-6} M	6	10^{-8} M	8	acidic

---

pH of Solutions

Calculate the pH of each solution.

1. 0.01 M HCl	pH = 2
2. 0.0010 M NaOH	pH = 11
3. 0.050 M $Ca(OH)_2$	pH = 1.3
4. 0.030 M HBr	pH = 1.5
5. 0.150 M KOH	pH = 13.2
6. 2.0 M $HC_2H_3O_2$ (Assume 5.0% dissociation.)	pH = 1.0
7. 3.0 M HF (Assume 10.0% dissociation.)	pH = 0.52
8. 0.50 M HNO_3	pH = 0.30
9. 2.50 M NH_4OH (Assume 5.00% dissociation.)	pH = 13.1
10. 5.0 M HNO_2 (Assume 1.0% dissociation.)	pH = 1.3

Acid-Base Titration

To determine the concentration of an acid (or base), we can react it with a base (or acid) of known concentration until it is completely neutralized. This point of exact neutralization, known as **endpoint**, is noted by the change in color of the indicator. Use the following equation:

$$N_A \times V_A = N_B \times V_B$$

where N = normality
V = volume

Solve each problem.

1. A 25.0 mL sample of HCl was titrated to the endpoint with 15.0 mL of 2.0 N NaOH. What was the normality of the HCl?

   1.2 N

2. A 10.0 mL sample of H_2SO_4 was exactly neutralized by 13.5 mL of 1.0 M KOH. What is the normality of the H_2SO_4?

   1.4 N

3. How much 1.5 M NaOH is necessary to exactly neutralize 20.0 mL of 2.5 M H_3PO_4?

   33 mL

4. How much 0.5 M HNO_3 is necessary to titrate 25.0 mL of 0.05 M $Ca(OH)_2$ solution to the endpoint?

   2.5 mL

5. What is the molarity of a NaOH solution if 15.0 mL is exactly neutralized by 7.5 mL of a 0.02 M $HC_2H_3O_2$ solution?

   0.01 M

Answer Key

Hydrolysis of Salts

Salt solutions may be acidic, basic, or neutral, depending on the original acid and base that formed the salt.

strong acid + strong base → neutral salt
strong acid + weak base → acidic salt
weak acid + strong base → basic salt

A weak acid and a weak base will produce any type of solution depending on the relative strengths of the acid and base involved.

Complete the chart for each salt shown.

	Salt	Parent Acid	Parent Base	Type of Solution
1.	KCl	HCl	KOH	neutral
2.	NH_4NO_3	HNO_3	NH_4OH ($NH_3 + H_2O$)	acidic
3.	Na_3PO_4	H_3PO_4	$NaOH$	basic
4.	$CaSO_4$	H_2SO_4	$Ca(OH)_2$	neutral
5.	$AlBr_3$	HBr	$Al(OH)_3$	acidic
6.	CuI_2	HI	$Cu(OH)_2$	acidic
7.	MgF_2	$Mg(OH)_2$	HF	basic
8.	$NaNO_3$	HNO_3	$NaOH$	neutral
9.	$LiC_2H_3O_2$	$HC_2H_3O_2$	$LiOH$	basic
10.	$ZnCl_2$	HCl	$Zn(OH)_2$	acidic
11.	$SrSO_4$	H_2SO_4	$Sr(OH)_2$	neutral
12.	$Ba_3(PO_4)_2$	H_3PO_4	$Ba(OH)_2$	basic

89

Acids and Bases Crossword

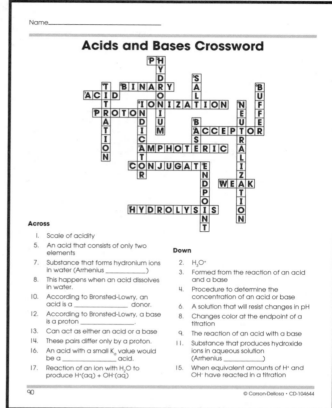

Across

1. Scale of acidity
5. An acid that consists of only two elements
7. Substance that forms hydronium ions in water (Arrhenius _____)
8. This happens when an acid dissolves in water.
10. According to Bronsted-Lowry, an acid is a _____ donor.
12. According to Bronsted-Lowry, a base is a proton _____.
13. Can act as either an acid or a base
14. These pairs differ only by a proton.
16. An acid with a small K_a value would be a _____ acid.
17. Reaction of an ion with H_2O to produce $H^+(aq)$ + $OH^-(aq)$

Down

2. H_3O^+
3. Formed from the reaction of an acid and a base
4. Procedure to determine the concentration of an acid or base
6. A solution that will resist changes in pH
8. Changes color at the endpoint of a titration
9. The reaction of an acid with a base
11. Substance that produces hydroxide ions in aqueous solution (Arrhenius _____)
15. When equivalent amounts of H^+ and OH^- have reacted in a titration

90

Assigning Oxidation Numbers

Assign oxidation numbers to all of the elements in each compound or ion shown.

1. HCl — $\overset{+1}{H}\overset{-1}{Cl}$
2. KNO_3 — $\overset{+1}{K}\overset{+5}{N}\overset{-2}{O_3}$
3. OH^- — $\overset{-2}{O}\overset{+1}{H^-}$
4. Mg_3N_2 — $\overset{+2}{Mg_3}\overset{-3}{N_2}$
5. $KClO_3$ — $\overset{+1}{K}\overset{+5}{Cl}\overset{-2}{O_3}$
6. $Al(NO_3)_3$ — $\overset{+3}{Al}\overset{+5}{(N}\overset{-2}{O_3)_3}$
7. S_8 — $\overset{0}{S_8}$
8. H_2O_2 — $\overset{+1}{H_2}\overset{-1}{O_2}$
9. PbO_2 — $\overset{+4}{Pb}\overset{-2}{O_2}$
10. $NaHSO_4$ — $\overset{+1}{Na}\overset{+1}{H}\overset{+6}{S}\overset{-2}{O_4}$

11. H_2SO_3 — $\overset{+1}{H_2}\overset{+4}{S}\overset{-2}{O_3}$
12. H_2SO_4 — $\overset{+1}{H_2}\overset{+6}{S}\overset{-2}{O_4}$
13. BaO_2 — $\overset{+2}{Ba}\overset{-1}{O_2}$
14. $KMnO_4$ — $\overset{+1}{K}\overset{+7}{Mn}\overset{-2}{O_4}$
15. LiH — $\overset{+1}{Li}\overset{-1}{H}$
16. MnO_2 — $\overset{+4}{Mn}\overset{-2}{O_2}$
17. OF_2 — $\overset{+2}{O}\overset{-1}{F_2}$
18. SO_3 — $\overset{+6}{S}\overset{-2}{O_3}$
19. NH_3 — $\overset{-3}{N}\overset{+1}{H_3}$
20. Na — $\overset{0}{Na}$

91

Redox Reactions

For each equation, identify the substance oxidized, the substance reduced, the oxidizing agent, and the reducing agent. Then, write the oxidation and reduction half-reactions.

Example:

$$\overset{\text{oxidized}}{Mg} + \overset{\text{reduced}}{Br_2} \to MgBr_2$$

reducing agent   oxidizing agent

oxidation half-reaction: $Mg^0 \to Mg^{+2} + 2e^-$
reduction half-reaction: $2e^- + Br_2^0 \to 2Br^-$

oxidized   reduced
1. $2H_2 + O_2 \to 2H_2O$
reducing agent   oxidizing agent

oxidation half-reaction: $2H_2^0 \to 4H^+ + 4e^-$
reduction half-reaction: $4e^- + O_2^0 \to 2O^{2-}$

oxidized   reduced
2. $Fe + Zn^{2+} \to Fe^{2+} + Zn$
reducing agent   oxidizing agent

oxidation half-reaction: $Fe \to Fe^{2+} + 2e^-$
reduction half-reaction: $2e^- + Zn^{2+} \to Zn^0$

oxidized   reduced
3. $2Al + 3Fe^{2+} \to 2Al^{3+} + 3Fe$
reducing agent   oxidizing agent

oxidation half-reaction: $2Al^0 \to 2Al^{3+} + 6e^-$
reduction half-reaction: $6e^- + 3Fe^{2+} \to 3Fe^0$

oxidized   reduced
4. $Cu + 2AgNO_3 \to Cu(NO_3)_2 + 2Ag$
reducing agent   oxidizing agent

oxidation half-reaction: $Cu \to Cu^{2+} + 2e^-$
reduction half-reaction: $2e^- + 2Ag^+ \to 2Ag$

92

Answer Key

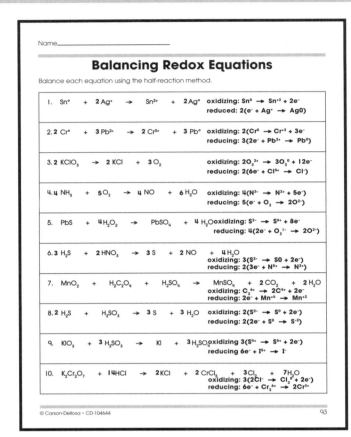

Balancing Redox Equations

Balance each equation using the half-reaction method.

1. $Sn^0 + 2Ag^+ \rightarrow Sn^{2+} + 2Ag^0$
   oxidizing: $Sn^0 \rightarrow Sn^{+2} + 2e^-$
   reduced: $2(e^- + Ag^+ \rightarrow Ag0)$

2. $2Cr^0 + 3Pb^{2+} \rightarrow 2Cr^{3+} + 3Pb^0$
   oxidizing: $2(Cr^0 \rightarrow Cr^{+3} + 3e^-)$
   reducing: $3(2e^- + Pb^{2+} \rightarrow Pb^0)$

3. $2KClO_3 \rightarrow 2KCl + 3O_2$
   oxidizing: $2O_3^{2+} \rightarrow 3O_2^0 + 12e^-$
   reducing: $2(6e^- + Cl^{5+} \rightarrow Cl^-)$

4. $4NH_3 + 5O_2 \rightarrow 4NO + 6H_2O$
   oxidizing: $4(N^{3-} \rightarrow N^{2+} + 5e^-)$
   reducing: $5(e^- + O_2 \rightarrow 2O^{2-})$

5. $PbS + 4H_2O_2 \rightarrow PbSO_4 + 4H_2O$
   oxidizing: $S^{2-} \rightarrow S^{6+} + 8e^-$
   reducing: $4(2e^- + O_2^{1-} \rightarrow 2O^{2-})$

6. $3H_2S + 2HNO_3 \rightarrow 3S + 2NO + 4H_2O$
   oxidizing: $3(S^{2-} \rightarrow S0 + 2e^-)$
   reducing: $2(3e^- + N^{5+} \rightarrow N^{2+})$

7. $MnO_2 + H_2C_2O_4 + H_2SO_4 \rightarrow MnSO_4 + 2CO_2 + 2H_2O$
   oxidizing: $C_2^{6+} \rightarrow 2C^{4+} + 2e^-$
   reducing: $2e^- + Mn^{+4} \rightarrow Mn^{+2}$

8. $2H_2S + H_2SO_3 \rightarrow 3S + 3H_2O$
   oxidizing: $2(S^{2-} \rightarrow S^0 + 2e^-)$
   reducing: $2(2e^- + S^0 \rightarrow S^{-2})$

9. $KIO_3 + 3H_2SO_3 \rightarrow KI + 3H_2SO_4$
   oxidizing: $3(S^{4+} \rightarrow S^{6+} + 2e^-)$
   reducing: $6e^- + I^{5+} \rightarrow I^-$

10. $K_2Cr_2O_7 + 14HCl \rightarrow 2KCl + 2CrCl_3 + 3Cl_2 + 7H_2O$
    oxidizing: $3(2Cl^- \rightarrow Cl_2^0 + 2e^-)$
    reducing: $6e^- + Cr_2^{6+} \rightarrow 2Cr^{3+}$

© Carson-Dellosa • CD-104644   93

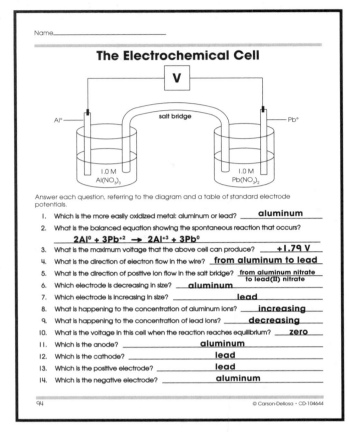

The Electrochemical Cell

Answer each question, referring to the diagram and a table of standard electrode potentials.

1. Which is the more easily oxidized metal: aluminum or lead? **aluminum**
2. What is the balanced equation showing the spontaneous reaction that occurs?
   $2Al^0 + 3Pb^{+2} \rightarrow 2Al^{+3} + 3Pb^0$
3. What is the maximum voltage that the above cell can produce? **+1.79 V**
4. What is the direction of electron flow in the wire? **from aluminum to lead**
5. What is the direction of positive ion flow in the salt bridge? **from aluminum nitrate to lead(II) nitrate**
6. Which electrode is decreasing in size? **aluminum**
7. Which electrode is increasing in size? **lead**
8. What is happening to the concentration of aluminum ions? **increasing**
9. What is happening to the concentration of lead ions? **decreasing**
10. What is the voltage in this cell when the reaction reaches equilibrium? **zero**
11. Which is the anode? **aluminum**
12. Which is the cathode? **lead**
13. Which is the positive electrode? **lead**
14. Which is the negative electrode? **aluminum**

94   © Carson-Dellosa • CD-104644

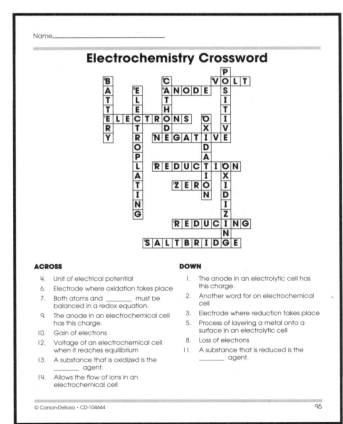

Electrochemistry Crossword

ACROSS

4. Unit of electrical potential
6. Electrode where oxidation takes place
7. Both atoms and _____ must be balanced in a redox equation.
9. The anode in an electrochemical cell has this charge.
10. Gain of electrons
12. Voltage of an electrochemical cell when it reaches equilibrium
13. A substance that is oxidized is the _____ agent.
14. Allows the flow of ions in an electrochemical cell

DOWN

1. The anode in an electrolytic cell has this charge.
2. Another word for on electrochemical cell
3. Electrode where reduction takes place
5. Process of layering a metal onto a surface in an electrolytic cell
8. Loss of electrons
11. A substance that is reduced is the _____ agent.

© Carson-Dellosa • CD-104644   95

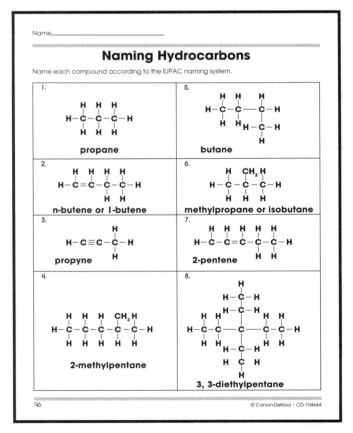

Naming Hydrocarbons

Name each compound according to the IUPAC naming system.

1. propane
2. n-butene or 1-butene
3. propyne
4. 2-methylpentane
5. butane
6. methylpropane or isobutane
7. 2-pentene
8. 3, 3-diethylpentane

96   © Carson-Dellosa • CD-104644

© Carson-Dellosa • CD-104644

Answer Key

Structure of Hydrocarbons

Draw the structure of each compound.

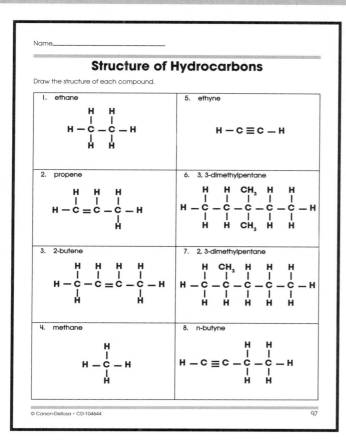

Functional Groups

Classify each of the organic compounds as an *alcohol, carboxylic acid, aldehyde, ketone, ether,* or *ester.* Then, draw its structural formula.

1. CH_3COOH **carboxylic acid**
2. CH_3COCH_3 **ketone**
3. CH_3CH_2OH **alcohol**
4. $CH_3CH_2OCH_3$ **ether**
5. CH_3CH_2CHO **aldehyde**
6. $CH_3CH(OH)CH_3$ **alcohol**
7. CH_3CH_2COOH **carboxylic acid**
8. $CH_3CH_2COOCH_3$ **ester**
9. $CH_3CH_2COCH_3$ **ketone**
10. CH_3OCH_3 **ether**

Naming Other Organic Compounds

Name each compound.

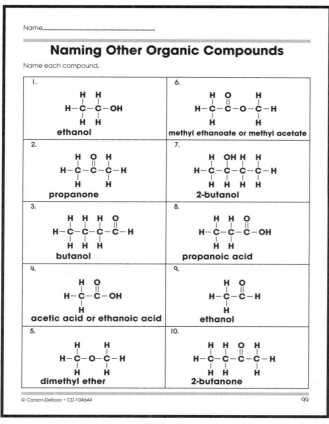

Structures of Other Organic Compounds

Draw the structure of each compound.

1. butanoic acid
2. methanal
3. methanol
4. butanone
5. diethyl ether
6. methyl methanoate (methyl formate)
7. 3-pentanol
8. methanoic acid (formic acid)
9. propanal
10. 2-pentanone

Answer Key

Organic Chemistry Crossword

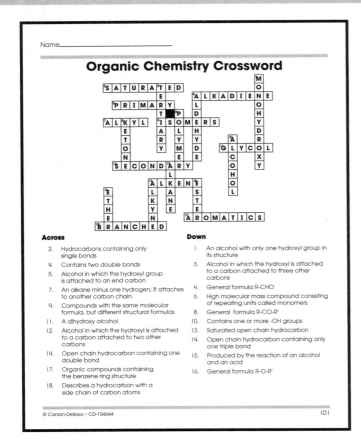

Across

2. Hydrocarbons containing only single bonds
4. Contains two double bonds
5. Alcohol in which the hydroxyl group is attached to an end carbon
7. An alkane minus one hydrogen; It attaches to another carbon chain.
9. Compounds with the same molecular formula, but different structural formulas
11. A dihydroxy alcohol
12. Alcohol in which the hydroxyl is attached to a carbon attached to two other carbons
14. Open chain hydrocarbon containing one double bond
17. Organic compounds containing the benzene ring structure
18. Describes a hydrocarbon with a side chain of carbon atoms

Down

1. An alcohol with only one hydroxyl group in its structure
3. Alcohol in which the hydroxyl is attached to a carbon attached to three other carbons
4. General formula R-CHO
6. High molecular mass compound consisting of repeating units called monomers
8. General formula R-CO-R'
10. Contains one or more -OH groups
13. Saturated open chain hydrocarbon
14. Open chain hydrocarbon containing only one triple bond
15. Produced by the reaction of an alcohol and an acid
16. General formula R-O-R'

101